# 1st International Conference on Energy Systems (ICES-2017)



Journal of Physics: Conference Series
Volume 1066

Belgorod, Russia
23 – 27 November 2017

ISBN: 978-1-5108-7200-4
ISSN: 1742-6588

**Printed from e-media with permission by:**

Curran Associates, Inc.
57 Morehouse Lane
Red Hook, NY 12571

**Some format issues inherent in the e-media version may also appear in this print version.**



Printed by Curran Associates, Inc. (2018)

For permission requests, please contact the Institute of Physics
at the address below.

Institute of Physics
Dirac House, Temple Back
Bristol BS1 6BE UK

Phone: 44 1 17 929 7481
Fax: 44 1 17 920 0979

techtracking@iop.org

**Additional copies of this publication are available from:**

Curran Associates, Inc.
57 Morehouse Lane
Red Hook, NY 12571 USA
Phone: 845-758-0400
Fax: 845-758-2633
Email: curran@proceedings.com
Web: www.proceedings.com

# Table of contents

## Volume 1066

**International Conference on Energy Systems-2017 (ICES-2017)**

**23–27 November 2017, Belgorod, Russian Federation**

**Accepted papers received: 20 July 2018**
**Published online: 7 September 2018**

### Preface



### Papers

IOP Conf. Series: Journal of Physics: Conf. Series **1066** (2018) 011001 doi:10.1088/1742-6596/1066/1/011001

# PREFACE. ICES 2017

The 1st conference of the International Conference on Energy Systems -2017 (ICES-2017) series, held in Belgorod, Russia from 23-27 November 2017.

This volume of proceedings from the conference provides an opportunity for readers to engage with a selection of best papers that were presented during the 1st ICES 2017 conference.

Divided into 9 sections, the 28 papers published here pick up on the multiple meanings of the Energy systems.

The Conference provided a setting for discussing recent developments in a wide variety of topics including the main engineering directions that are Heat power, Electrical Power, Power Engineering, Power supply systems, Energy saving and energy efficiency, Alternative and renewable energy sources, Ecology of power engineering, Heating and Building Thermal Physics, Energy resource saving in technologies.

One of the unique and valuable dimensions to the ICES conference series is the way the series brings engineering educators together from around the world and different departments to discuss ways to serve learners better. The Conference has been a good opportunity for participants coming from Ukraine, Iraq, Serbia to present and discuss topics in their respective research areas.

The reader will discover that both common challenges and creative solutions emerge from very diverse settings – examples include the Belgorod State technology university named after V.G. Shukhov (Russia), LTD "Center for Energy-Saving Technologies" (Belgorod, Russia), Ivanovo State Energy University (Ivanovo, Russia), National Research University "Moscow Power Engineering Institute" , Donbass Architectural and Construction Academy (Ukraine), University of Niš (Serbia), University of Diyala (Iraq).

We would like to thank all participants for their contributions to the Conference program and for their contributions to these Proceedings.

We are looking forward to the Second International Conference on Energy systems, that will be held on Fall 2018 at the same location. We hope that it will be an interesting and enjoying at least as all of its twenty nine predecessors.

**Alexander Belousov**-Chairman of the ICES-2017 conference
Cand. Tech. Sciences, Associate Professor, Director of the Institute of Energy, Information Technology and Control Systems, Head of Electric power engineering and automation Department of BSTU named after V.G. Shukhov

**Pavel Trubaev** - Deputy of Chairman of the ICES-2017 conference
Dr. Tech. Sci., Professor of the Department of Energy of Heat Technologies of BSTU named after V.G. Shukhov

# Peer review statement

All papers published in this volume of *Journal of Physics: Conference Series* have been peer reviewed through processes administered by the proceedings Editors. Reviews were conducted by expert referees to the professional and scientific standards expected of a proceedings journal published by IOP Publishing.

IOP Conf. Series: Journal of Physics: Conf. Series **1066** (2018) 012001  doi:10.1088/1742-6596/1066/1/012001

# ALGORITHMS OF THE MINIMUM SPANNING TREE IN GAS SUPPLY SYSTEMS

**D O Temnikov and D Yu Suslov**

Department of Heat and Gas Supply and Ventilation, Belgorod State Technological University of V.G. Shukhov, Belgorod, 308012, RU

**Abstract.** In this paper, a method for tracing gas networks using the Kruskal algorithm and the Steiner algorithm is considered. The result of the work of algorithms on the example of a bio meta network between agricultural enterprises located on the territory of the Belgorod region is demonstrated. The economic result is determined using Steiner's refinement algorithm.

At present, when designing gas supply systems, the actual task is to use automatic network tracing. Automatic tracing allows you to save time and money both for designing and building networks, as the task of automatic routing is to reduce the length of pipes [1].

The automatic trace process is divided into two phases: global tracing and detailed tracing. At the stage of global tracing, a preliminary route is created based on the unit cost of the pipe meter or the cost of its laying, as well as the total length of the entire route [2].

The purpose of this paper is to determine the economically feasible method of designing gas networks by comparing the result of automatic routing using the algorithms of Kruskal and Steiner.

To achieve this goal, it is necessary to adapt the algorithms and compare the total length.

The construction of the minimal spanning tree is solved by algorithmization and has two principal differences.

1. Construction without additional vertices:
- Prima-Kraskal algorithm;
- The Borovka algorithm;
- Dijkstra's algorithm.

2. Construction with additional vertices:
- Steiner trees (Euclidean plane);
- Steiner trees (Rectangular plane);
- Melzak's algorithm;
- Cocaine's algorithm.

The Steiner algorithm for the Euclidean plane can find its application in gas supply by bio methane between agricultural enterprises or for tracing inter-settlement or magistral gas pipelines where there are no hard-coded security zones for laying, such as in a city.

The Steiner algorithm for a rectangular problem can be used for intra-quarter gas distribution networks, for which a rectangular topology is typical and the designation zones are rigidly identified.

E. Gilbert and H. Pollack showed that the Steiner tree can reduce the minimum spanning tree to 13.4% [3].

The Steiner algorithm has four properties:

1. The number of edges of a tree is one less than the total number of initial vertices and Steiner points.
2. Any two edges emerging from the same vertex form an angle of at least 120 °.
3. One, two or three edges can emerge from this vertex; if two edges come out, then the angle between them is greater than or equal to 120 °; if three, then they form angles of 120 ° with each other.
4. Three edges with angles of 120 ° emerge from each additional vertex (Figure 1).

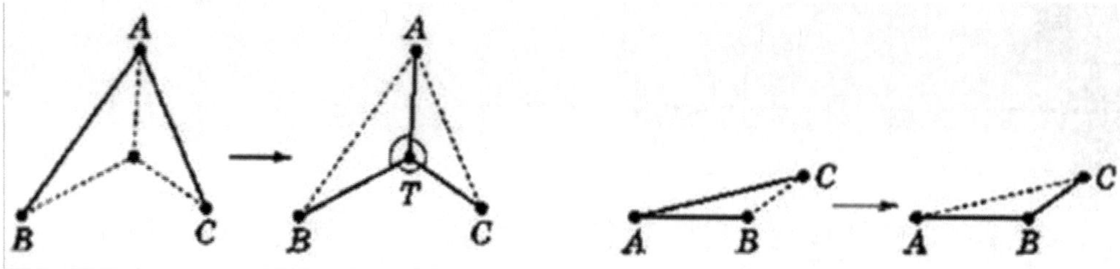


**Figure 1.** Stages of operation of the Steiner algorithm

The Steiner algorithm based on the starting points builds a set of different trees that satisfy the above conditions. Some of these trees are called locally minimal trees. Not every locally minimal tree is the shortest possible solution of the problem. To transform the network into the shortest tree, called the globally minimal Steiner tree, global changes in the displacement of points can be used. As a result, there is a complete search of all possible locally minimal Steiner trees to find the shortest path (Figure 2). Since there is a constant change in the location of the Steiner points, the time to create all the minimal local trees and their search can be unlimited.

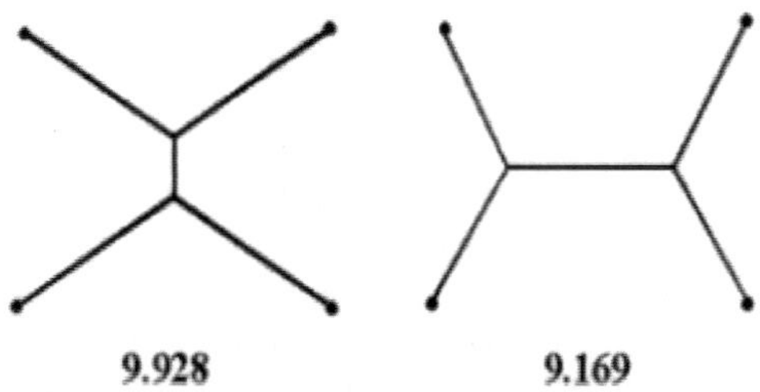


**Figure 2.** Relatively minimal Steiner trees

Further development of the Steiner algorithm was obtained by Melzak, who divided the set of all points into subsets and applied the Steiner algorithm in each subset.

The Steiner problem is NP-complete, so the algorithms of this problem can not be used in CAD because of the long calculation time. As a result, heuristic algorithms were developed. One successful algorithm is the sequential introduction of additional vertices into the Prima-Kraskal tree [4].

We apply the Kruskal algorithm and Steiner algorithm for designing a biomethane gas pipeline between agricultural enterprises in the Belgorod region (Figure 3).

The total area of the territory of the Belgorod region is 27,100 km2. There are 136 agricultural enterprises on the territory. We shall regard them as nodes of the graph.

We construct the minimal spanning tree using the Kruskal algorithm (Fig. 4). The length of the constructed network is 2 626.31 km.

We construct the Steiner tree on the basis of the Kruskal algorithm (Fig. 5). The length of Steiner's tree is 2 548.8 km, which is 77.5 km less or 2.95%.

In monetary terms, savings on working documentation during design is 37.66 million rubles and 413.51 million rubles at work [5].

Using the algorithm of the minimal covering spanning tree, a bio metric gas network was built between the enterprises of the agro-industrial complex of the Belgorod region. Then the bio metric network was refined using the Steiner algorithm.

It can be concluded that the length of the gas network constructed according to the Steiner algorithm is 77.5 km shorter. Thus, the savings on construction of the network amounted to 451.17 million rubles.

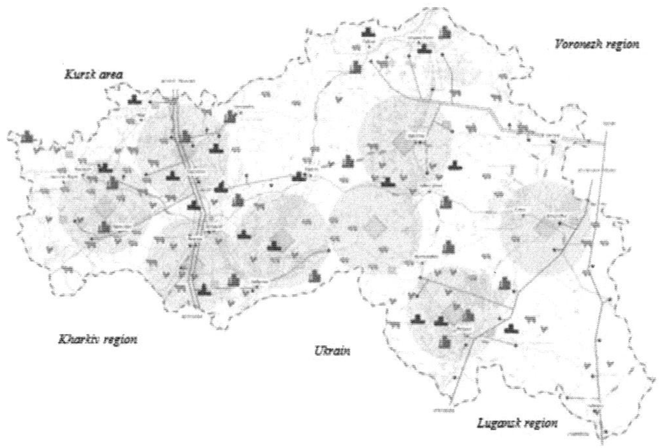


**Figure 3.** Location of agricultural enterprises in the territory Belgorod region

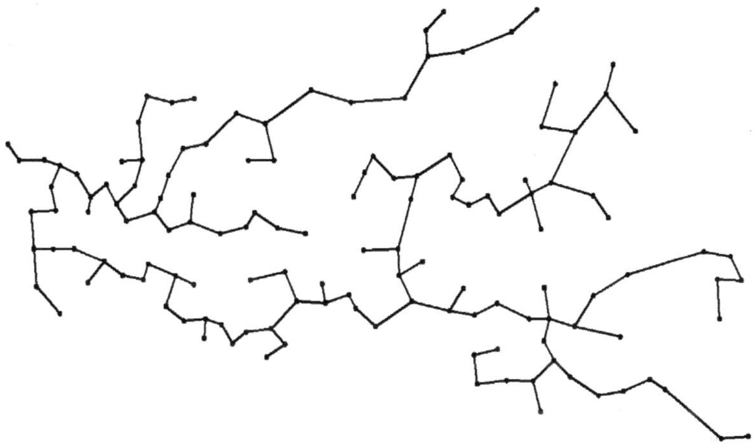

**Figure 4.** Minimal spanning tree Kraskala

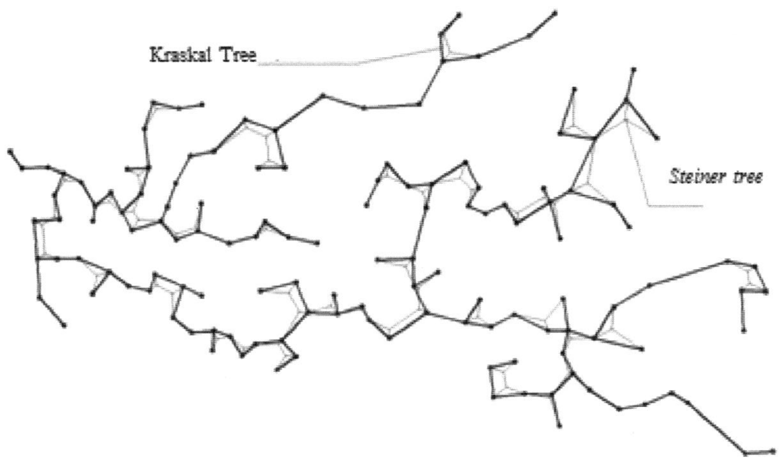


**Figure 5.** The Kraskal Tree and the Steiner Tree

**Acknowledgments**
The article was prepared within development program of the Flagship Regional University on the basis of Belgorod State Technological University named after V.G. Shukhov.

**References**

[1] Suslov D, Vyrodov G 2016 Gasdynamic calculation of gas networks of low pressure with the use of CAD *Bulletin of Belgorod State Technological University named after V G Shukhov* vol 10 pp 114-119

[1] Ryzhenko N 2003 The problem of constructing a Steiner tree for the global CAD trace phase *High-performance computing systems and microprocessors: Proceedings of IMVS RAS* vol 4 pp 96-105

[2] Gilbert E and Pollak H 1968 *SIAM J. on App. Math.* **16-1** 1-29

[3] Barrera T, Griffith J, McKee S 1993 Toward a Steiner engine: enhanced and parallel implementations of the iterated 1-Steiner MRST algorithm. *Proc. Great Lakes Symposium on VLSI* (Kalamazoo MI) pp 90–94

[4] SBTSP 81-02-14-2001 State Estimated Standard "Reference book for basic prices for design work in construction. Gas equipment and gas supply of industrial enterprises, buildings and structures"

IOP Conf. Series: Journal of Physics: Conf. Series **1066** (2018) 012002 doi:10.1088/1742-6596/1066/1/012002

# ANALYSIS OF DESIGN FEATURES OF CIRCULATING ACTION MIXERS

**M V Sevostyanov , V S Sevostyanov and I G Martakov**

Department of Technological Complexes of Machines and Mechanisms, Belgorod State Technological University of V.G. Shukhov, Belgorod, 308012, RU

**Abstract**. This article considers the classification of the mixers of the circulation action. The analysis of their structural peculiarities and disadvantages was carried out. The need to create combined mixing devices has been identified. The design of a combined mixing apparatus is proposed.

One of the most important technological stages of most production is the process of mixing various bulk and viscous components. This is typical for various branches of industries such as the industry of building materials, food, metallurgy, pharmaceutical, agro chemistry and agriculture, chemical and etc. industries. The main purpose of the mixing process is to obtain a homogeneous mixture from separate components.

In the process of mixing bulk materials that differ in the density or particle size, the segregation, separation of particles by their size or density, affects the quality of the finished mixture negatively.

One of the effective methods of counteracting this process is the ordered loading of the components of the mixture and its homogenization in a circulating mixer [1, 2].

At the same time, segregation cannot prevent, but promote the production of high-quality mixtures. This is due to the fact that circulating mixers have an ordered nature of the motion of the components (the motion of material flows through a closed contour of varying complexity) and the zones of segregated state of the mixture are easily predictable.

The aim of this paper is to analyze the designs of the mixers of the circulation action and to develop a mixer design of the combined action.

Among circulating-type mixers, was planetary-screw, bladed, spinning centrifugal and screw with a central circulation pipe became more distributed in industry. They are recommended for mixing lightweight disconnected and cohesive bulk materials, as well as for mixing bulk materials with a small amount of liquid.

A mixer with a planetary-screw agitator [3] consists of the following main units (Figure 1): conical housing 1; transmission boxes 2; augers 3; drive 4.

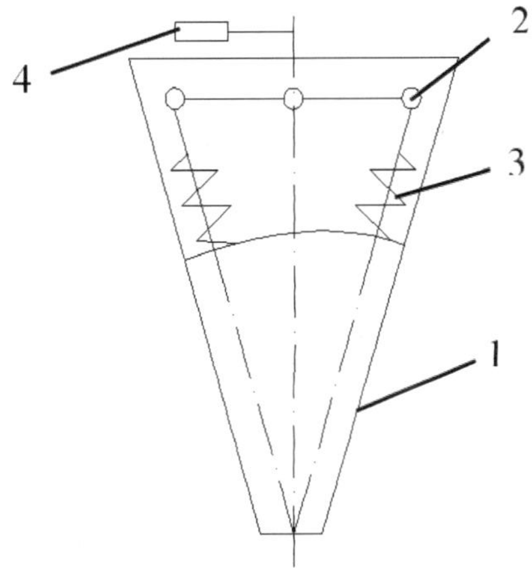


**Figure 1**. Mixer planetary-screw:
1 - conical housing; 2 – transmission box; 3 - screws; 4 – drive

The disadvantages of such mixers are the complexity of the design, due to the large number of gears, and the presence of a stagnant zone in the central parts of the mixer.

Blade mixers [4] consist of the following main units (Figure 2): a vertical cylindrical body 1; unloading device 2; working body 3; drive 4.

Disadvantages of such mixers are increased energy costs and stagnant zones, depending on the type of working element.

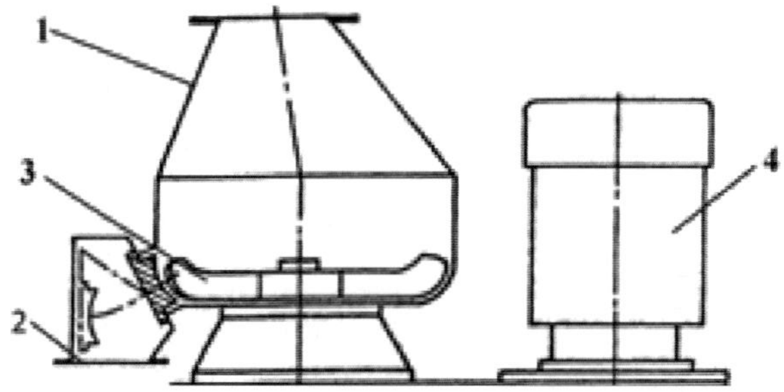


**Figure 2.** Blade mixer:
1 - vertical cylindrical body; 2 - unloading device; 3 - working body; 4 - drive

The spindle centrifugal mixer [5] consists of the following main units (Figure 3): cylinder-cone body 1; drive vertical shaft 2 with rotor 3; a hollow truncated cone 4 and a radial mixer 5; blade 6; drive 7.

The disadvantages of such mixers are increased energy costs and the complexity of the rotor design with blades.

A mixer with a central circulation pipe [6] consists of the following main parts (Fig. 4): a vertical body 1; with a vertical circulation pipe 2 installed in it; gas injection pipes 3; nozzles 4; unloading device 5; feed nozzle 6 and the air filter 7.

The disadvantages of such mixers are the complexity of the design, due to a gas supply and filtration system, and stagnant zones in the side parts of the mixer

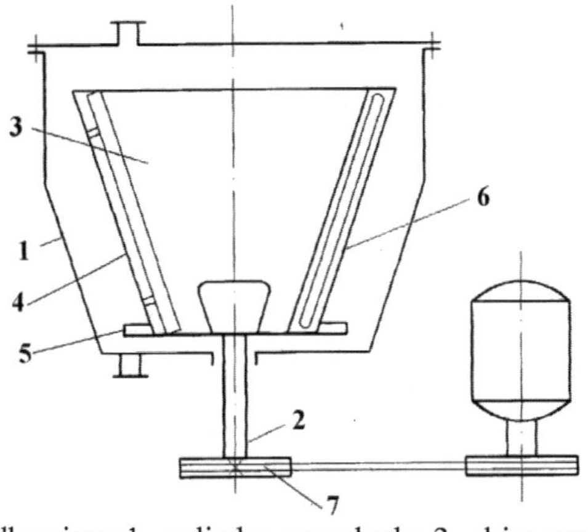


**Figure 3.** Centrifugal spindle mixer: 1 - cylinder- cone body; 2 - drive vertical shaft; 3 - rotor; 4 - hollow truncated cone; 5 - radial stirrer; 6 - blades; 7 – drive

An analysis of the existing constructions has shown that the working bodies of the mixers creating the circulation movement are mixers made in the form of augers, radial blades, propellers or disks.

The implementation of the mixing process using these structures does not guarantee obtaining a qualitative mixture without providing certain operating modes.

The solution of this problem is the creation of combined mixing plants, which allow combining different modes of operation of mixers with different types of working bodies and also provide the possibility of creating various turbulent-laminar flows and obtaining qualitative homogenized mixtures with different physical and mechanical characteristics of the incoming components.

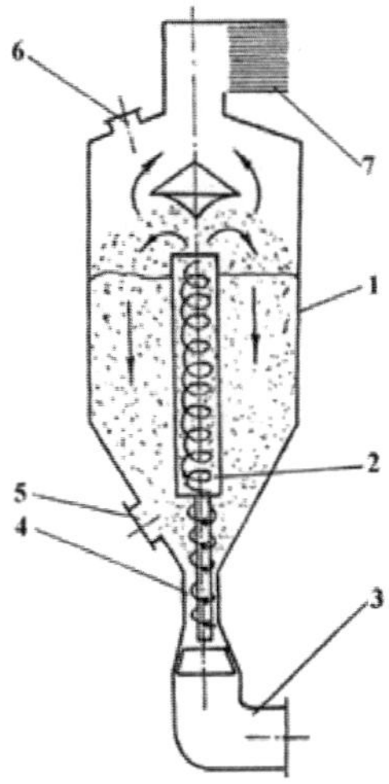


**Figure 4**. Mixer with central circulation pipe:
1 - vertical body; 2 - vertical circulation pipe; 3 - gas injection pipe; 4 - nozzle; 5 - unloading device; 6 - loading branch pipe; 7 - the air filter

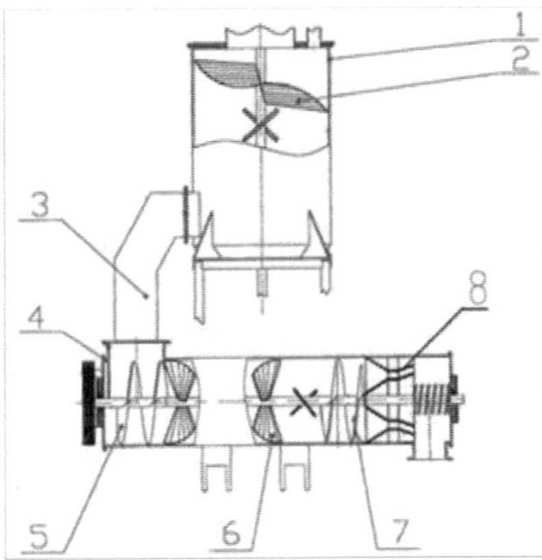


**Figure 5**. Technological mixing module: 1 - vertical mixer; 2 - unidirectional double-threaded screw blades; 3 - a branch pipe; 4 - horizontal mixer; 5, 7 – screw blades; 6 - oppositely directed double-threaded screw blades; 8 - block for mechanical pre-compaction of the mixture

In the developed technological module [7] a combined mixing of materials is carried out (Fig. 5). The realization of mixing techno genic fibrous materials with various physical and mechanical characteristics is due to the stage-by-stage high-speed mixing of the mixture with the organization of internal recycling at each stage of their mixing, and the consequent increase in its density through mechanical pre-compacting and micro granulation. Due to the consequently installed vertical and horizontal mixers with blades, mixing takes place in two stages. Turbulent-gyration mixing takes place in the first stage and in the second stage, recirculation is carried out with steam humidification. As a result of the analysis, it was found that the designs of the circulation mixers do not provide the required quality of the finished mixture. The use of the proposed mixer design combined action makes it possible to obtain ready-mixes of the required quality with different physical mechanical characteristics of the incoming components.

**Acknowledgments**
The article was prepared within development program of the Flagship Regional University on the basis of Belgorod State Technological University named after V.G. Shukhov.

**References**

[1] Selivanov Yu T, Pershin V F 2011 Some aspects of practical use of circulating mixers Chemical industry today **2** pp 51-56

IOP Conf. Series: Journal of Physics: Conf. Series **1066** (2018) 012002 doi:10.1088/1742-6596/1066/1/012002

[2] Glagolev S N, Sevostyanov V S, Gridchin A M and others 2013 Resource-energy-saving modules for the integrated utilization of techno genic materials *Vestnik of the Belgorod State Technological University named after V. G. Shukhov* **6** pp 102-106

[3] Orlov G I, Kupriyanov N A and Batashova Yu V Author's certificate of the USSR № 1011220, IPC B01F 7/02. Planetary- screw mixer (USSR) No 3 359369 / 23-26; claimed by 05.11.81; publ. 15.04.83 Bul 14 p 3

[4] Rastegaev P S, Burmaka V L and Manivchuk S D Author's certificate of the USSR No 1494957, IPC B01F 7/18 Centrifugal blade mixer / No 4116079 / 23-26; claimed by 01.07.86; publ. 23.07.89 Bul 27 p 2

[5] Borshchev V Ya The patent of the Russian Federation. No. 97935, IPC D03J 3/00. Spindle centrifugal mixer /; the applicant and the patent holder of Tambov State Technological University. No. 2010114959/05; claimed by 14.04.2010; publ. 27.09.2010 p 6

[6] Fiyalka M D Author's certificate of the USSR No. 1816490, IPC B01F 5/24. Pneumatic mixer / (USSR). No. 4921336/26; claimed by 01.02.91; publ. 23.05.93, Bul. No. 19. p 3

[7] Sevostyanov V S, Glagolev S N, Gridchin A M and others; The patent of the Russian Federation № 2624306, МПК B01F 7/04 (2006.01), B01F 7/18 (2006.01), B01F 13/10 (2006.01) 07/03/2017. Technological module and method of mixing technogenic fibrous materials the applicant and the patent holder Belgorod State Technological University. 2016140757; claimed by 10.17.2016; publ. 07.03.2017

IOP Conf. Series: Journal of Physics: Conf. Series **1066** (2018) 012003 doi:10.1088/1742-6596/1066/1/012003

# ANALYSIS OF MSW COMBUSTION TEMPERATURE IN A HOT WATER BOILER WITH THE LOW-CAPACITY

**N V Kornilova[1] and P A Trubaev[2]**

[1]Department of Power Engineering and Automation , Belgorod State TechnologicalUniversity of V.G. Shukhov, Belgorod, 308012, RU
[2]Department of Heat Engineering Technology ,Belgorod State TechnologicalUniversity of V.G. Shukhov, Belgorod, 308012, RU

**Abstract.** The paper analyzes the combustion temperature of solid waste in a small-capacity hot-water boiler with a great depending on the humidity of the solid waste, the excess air factor, the temperature of the air supplied to the combustion and the amount of heat taken and the range of parameters that provide the required combustion temperature is 900 ... 1100 ° C, which provides complete destruction of organic waste substances in the waste, but does not lead to ash melt and the formation of sprays. It is concluded that the necessary temperatures are provided for a wide range of regime parameters and the humidity of solid waste. But selection of heat in the boiler leads to a significant decrease in the combustion temperature. Therefore, in boilers for burning low-power MSW, it is necessary to constructively separate the combustion zone and the heat extraction.

At present, there is no cost-effective and simultaneously environmentally safe solution for the disposal of solid domestic waste (MSW) [1, 2]. The existing system of waste management in Russia is focused mainly on their disposal at landfills. This leads to the withdrawal from the use of large land areas, pollution of the atmosphere by bio gas and toxic combustion products in self-ignition of waste, and soil and water bodies - a toxic liquid ("filtrate") formed in the waste layer [3].

Incineration of domestic garbage is the most common and widespread method of its utilization. There are about 1500 incineration plants in the world, of which about one third - in Europe, where about 16% of waste is burned, mainly with the generation of thermal energy and slightly - with the generation of electricity [4]. In the US, the volume of MSW burned after the 2000s was comparable to the amount of waste placed at landfills. Burning domestic waste, in addition to reducing the volume and weight, allows you to get additional energy resources that can be used for centralized heating and power generation. In addition, the storage of the remaining part of solid waste after burning is environmentally safe and requires 10 - 12 times less area.

There are three main methods for burning solid waste - layered (with a fixed and moving grate or chain grate); pulverized (in the suspended or boiling layer) and pyrolysis boilers. The first method is easier to implement, does not require preliminary preparation of debris, is highly reliable, but the second allows for more complete combustion of waste. According to the materials of a number of sources, pyrolysis is the most economically effective and has the least impact on the environment [6]. However, the opinions of specialists differ on the effectiveness of dry pyrolysis in the combustion of solid domestic and certain industrial wastes [7].

The main disadvantage of direct combustion is pollution of the atmosphere by harmful emissions with an excess of oxygen in the combustion zone and a low combustion temperature [8]. But for example, in work [7] it is stated that dumping of waste in landfills is more dangerous than processing by burning. The experience of Sweden shows that, despite the growth in the volume of domestic waste in recent years, the release of dioxins into the atmosphere from incineration plants is only 5-6% of all emissions, i.e. as much as in previous fires in landfills [4].

Thermal waste neutralization at the modern level of development of science and technology guarantees almost complete destruction of organic harmful substances in waste [8]. But for this it is necessary to ensure high temperatures. According to the Directive 2000/76 / EC of the European Parliament and of the

Council "On incineration of waste", installations in which the combustion products are at least 2 s at a temperature of at least 850 ° C or if dangerous waste containing more than 1% of halogen organic compounds expressed as chlorine, the temperature should be at least 1100 ° C.

But at temperatures above 1100 ° C, the ash produced by incineration of solid waste begins to melt, which leads to the formation of cakes [1], which clog the cells of the grate. Therefore, the temperature of combustion of solid waste should lie in the interval 850-1100 ° C.

The paper analyzes the combustion temperature of solid domestic waste in a hot-water boiler with a grate depending on the humidity of the solid waste, the excess air factor, the temperature of the air supplied to the combustion and the amount of heat taken and the range of parameters that provide the required combustion temperature is 900 ... 1100 ° For calculations, the composition of solid waste was used according to the data of [1, 9] (Table 1).

**Table 1. MSW composition**

| MSWcomposition | Share | Characteristicsofperformance, % | | | | | | |
|---|---|---|---|---|---|---|---|---|
| | | $A^p$ | $W^p$ | $C^p$ | $H^p$ | $O^p$ | $N^p$ | $S^p$ |
| Foodwaste | 0,3 | 3 | 76 | 11,26 | 1,62 | 7,16 | 0,84 | 0,13 |
| Paper | 0,3 | 12 | 2 | 39,73 | 5,33 | 40,51 | 0,26 | 0,17 |
| Tree | 0,05 | 1 | 67 | 16,32 | 1,95 | 13,63 | 0,06 | 0,03 |
| Plastic | 0,25 | 21 | 2 | 52,13 | 7,16 | 16,56 | 0,85 | 0,31 |
| Noncombustiblepart | 0,1 | 100 | 0 | - | - | - | - | - |
| Working composition of solid waste | | 19,8 | 27,25 | 29,14 | 3,97 | 19,12 | 0,54 | 0,17 |

The calculation was carried out for a 200 kW boiler incinerated at Ltd TK Ekontras (Belgorod, Serafimovich Str., 72) and designed to generate hot water for heating the administrative and industrial building. This solid-fueled boiler represents an all-welded construction of steel with dimensions of 1230 m (width) x 1600 mm (depth x 2170 mm (height).) Fuel is burned on the grate, the air intake can be either natural or blowing fan. The walls of the boiler are warmly insulated with mineral wool Izovol ST-40, the thickness of insulation is 0.2 m.

The boiler's heat balance was calculated, which includes the following articles:

– the arrival of heat: a) from the combustion of fuel; b) physical heat of fuel; c) heat of air;

– heat consumption: a) the heat of the combustion products (CO2, SO2, H2O, N2, O2); b) heat of ash; c) losses through the walls of the boiler; d) losses from under burning; e) heat sink (the amount of heat taken in the heat exchanger for heating water).

According to the obtained heat of combustion products, their temperature was determined, which was assumed equal to the actual combustion temperature.

In Fig. 1 shows the dependence of the combustion temperature on the coefficient of excess air and humidity of solid waste at different temperatures of air supplied for combustion and the absence of heat in the boiler. In Fig. 2, from the data obtained, isotherms are constructed that determine the region of required temperatures. According to the graphs, MSW can be burned at a humidity of up to 20-30 without preheating the combustion air. Higher humidity of solid waste requires the use of preheating of air, which can be carried out due to the heat of the combustion products.

Unlike large-capacity boilers, where the combustion and heat dissipation zones are spatially separated, in boilers of small production, heat extraction takes place directly in the combustion chamber, which leads to a decrease in the combustion temperature. The calculation of the combustion temperature in the presence of heat removal in the boiler showed , that even an insignificant part of the heat, taken in the furnace, leads to the impossibility of obtaining the required temperatures (Fig. 3).

IOP Conf. Series: Journal of Physics: Conf. Series **1066** (2018) 012003 doi:10.1088/1742-6596/1066/1/012003

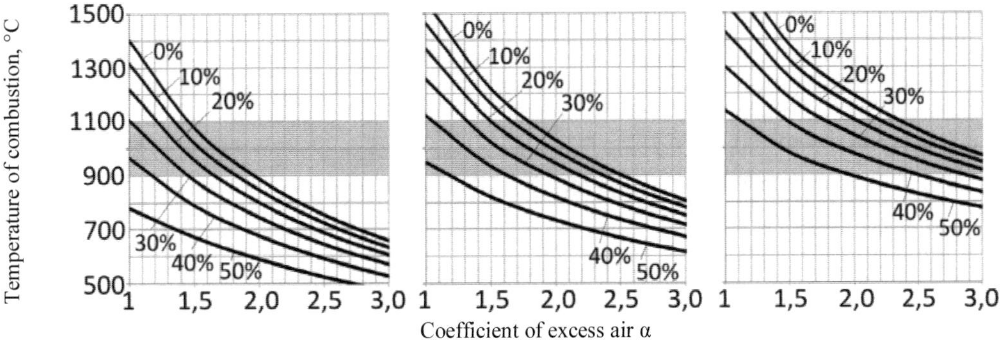


a) b) c)

**Figure 1.** The combustion temperature for different humidity of solid waste and the temperature of the air supplied for combustion: a) 20 ° C; b) 200 ° C; c) 400 ° C

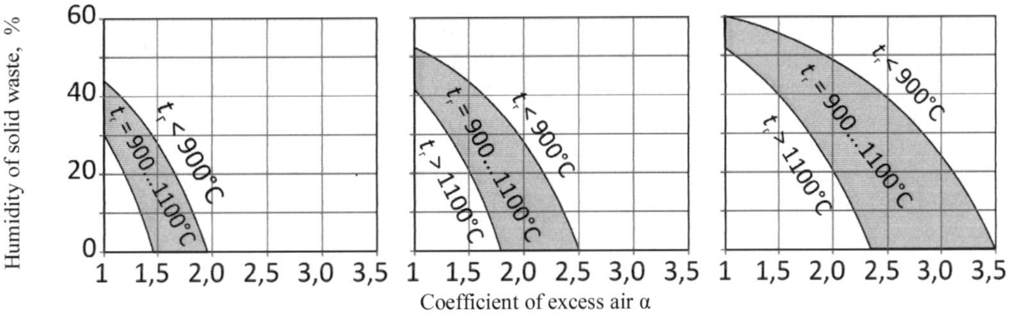


a) b) c)

**Figure 2.** Isotherms of the combustion temperature of solid waste at the temperature of the air supplied for combustion: a) 20 ° C; b) 200 ° C; c) 400 ° C

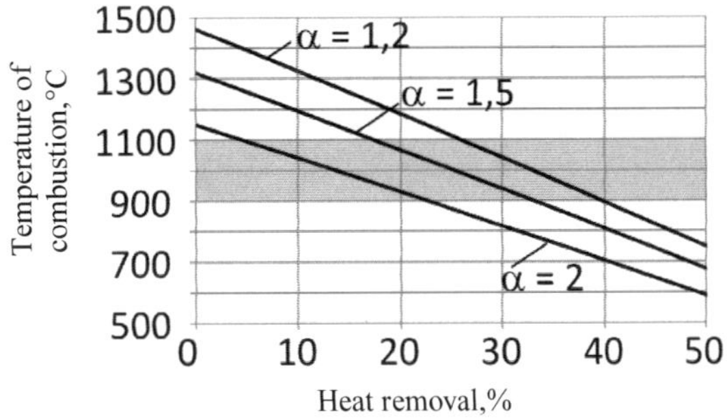


**Figure 3**. Change in the combustion temperature in the boiler for different heat removal (heat removal)

The obtained data are consistent with the results of the boiler tests during incineration, in which the pyrometer measured the temperature of the fuel layer in the boiler ($t_g$), and the gas analyzer - the composition and temperature of the gas ($t_g$) after the boiler (Fig. 4). The boiler output was about 50 kW, or about 25% of the heat removal.

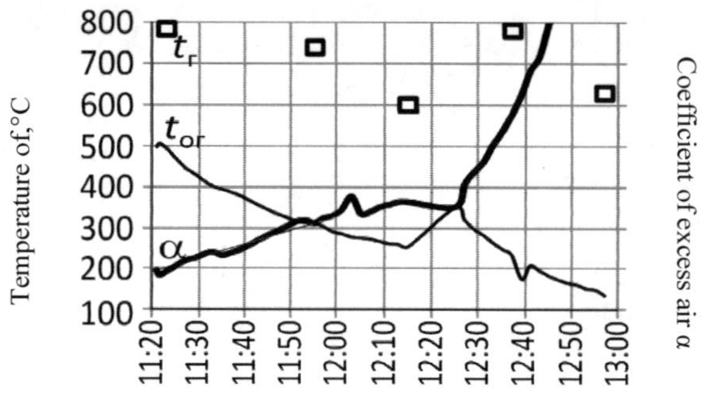


**Figure 4.** The data of the measurements of the combustion temperature t, the gases leaving the furnace and the heat of the excess air with a heat dissipation in the boiler of about 25%

Thus, despite the relatively high combustion temperature, the temperature of the exhaust gases is far below the values determined by environmental requirements. Although the measurements of the off-gas composition given in [8] showed no exceeding of the MPC for the analyzed gases, but according to the results of the tests it can be concluded that the environmental requirements call for the required combustion regimes in the boiler in the absence of heat extraction for heating the water.

To conclude it should be said thatwhen MSW is incinerated, the necessary temperatures can be obtained that ensure complete combustion of harmful substances, with a wide range of regime parameters and humidity. But selection of heat in the boiler leads to a significant decrease in the combustion temperature. Therefore, in boilers for burning low-power MSW, it is necessary to constructively separate the combustion zone and the heat extraction. Taking into account the requirements for the presence of combustion products for at least 2 s at a temperature of at least 850 ° C, the furnace and the heat exchanger must be separated by a gas flue of the required size.

**Acknowledgments**
The article was prepared within development program of the Flagship Regional University on the basis of Belgorod State Technological University named after V.G. Shukhov.

**References**
[1] Levin B 1982 Use of solid domestic waste in energy supply systems (Moscow: Energoizdat ) p 224
[2] Kozhevnikov V, Tokach Yu, Ognev M 2015 Modern solutions for processing of solid domestic waste in BSTU V G Shukhova *Bulletin of the Belgorod State Technological University named after V G Shukhov*vol 1 pp 172-174
[3] Porozhnyuk L, Vasilenko T, Porozhnyuk E 2012 Role of environmental audit in handling waste in the Belgorod region Bulletin *of the Belgorod State Technology University named after V G Shukhov*vol 4 pp 177-180
[4] European practice of waste management: problems, solutions, prospects 2004 (S Pb : NP REP 2004)p 73
[5] Tugov A 2014 Perspectives of energy utilization of solid waste (Energosovet) vol 4 pp 31-35
[6] Korovin I 2003 Study pyrolysis utilization of carbon-containing solid household waste *Dissertaion of Candidat of tech. Sciences* ( Tyumen 2003) p 159
[7] Kasakura T, Hiraoka M 1982 Pilot plant study on sewage sludge pyrolysis Water Research **16** pp 1335-1348, 1569-1575
[8] Kornilova N, Trubayev P 2016 Determination of harmful emissions of a water-and-gas boiler for the incineration of solid wastes of various types. Interna- tional Scientific and Research Journal pp 10-2, 62-66
[9] Ilyinikh G, Slyusar N, Korotaev V 2011 Morphological composition of waste: the main trends of change. Solid household waste vol 8 pp 38-41

# APPLICATION OF BIOMETHANE FOR GAS SUPPLY WITHIN THE SETTLEMENTS

**D Y Suslov**

Department of Heat and Gas Supply and Ventilation, BelgorodState Technological University of V.G. Shukhov, Belgorod, 308012, RU

**Abstract**. The work is devoted to the problem of using bio gas fuel in gas supply systems. The analysis of the work of bio gas stations and bio methane plants in the countries of the European Union was conducted, the ways of using the produced bio gas were considered. The possibility and prospects of using bio gas in centralized gas supply systems have been studied. It is established that the total potential of bio methane production from agricultural waste of the Belgorod region is 465 million m3 / year, which is 14% of the annual volume of gas consumption in the region.

Recently, in developed countries, there has been a steady trend towards reducing the consumption of exhaustible energy sources and using alternative fuels. In most countries, state programs and laws have been adopted and are working to develop small-scale power generation and use of environmentally friendly renewable energy sources [1].

In the countries of the European Union (EU), the Unified Environmental Strategy till 2020 "Strategy 20-20-20" was adopted. According to this strategy, by 2020, the level of carbon dioxide emissions to the atmosphere should be reduced by 20% (compared to the level of 1999), the share of energy from renewable sources in the total energy consumption structure - grow up to 20%, and the total energy consumption - to be reduced by 20% [ 1]. One of the important sectors of renewable energy in the world is the production and energy use of bio gas. The share of bio gas in the total balance of energy consumption in Europe is 13%.

The aim of the work is to assess the feasibility and prospects of using bio gas in centralized gas supply systems.

Bio gas is a gaseous energy source, obtained as a result of anaerobic fermentation of organic substances of various origin and composition. Bio gas consists mainly of methane ($CH_4$) and carbon dioxide ($CO_2$), but also contains a small amount of other gases: hydrogen sulfide ($H_2S$), nitrogen ($N_2$), hydrogen ($H_2$) [2].

The total amount of energy received from bio gas in the EU countries in 2015 was 15.612 million toe. Most of this energy (77%) was produced in three countries: Germany (7.9), Great Britain (2.3) and Italy (1.9 million toe) [3].

Currently, there are 17,240 bio gas plants operating in Europe (Figure 1) [3]. At most stations, the bio gas produced is used in cogeneration plants for the production of heat and electric energy with its supply to centralized electric networks. At the same time, a small part of the heat energy is used for needs of the bio gas plant, and the rest of the heat is not used and is utilized. This leads to a decrease in the economic efficiency of the bio gas plant and an increase in the payback period.

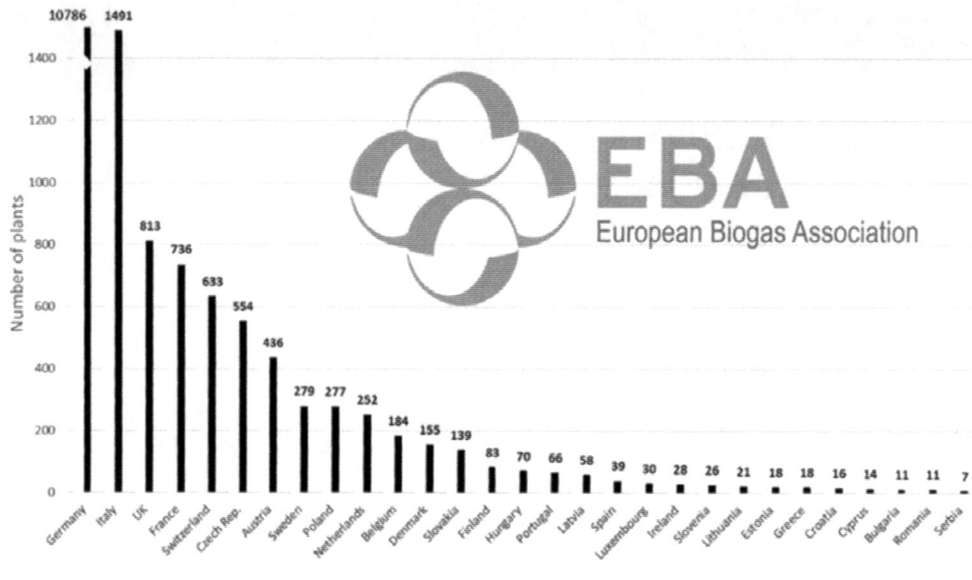


**Figure 1.** Number of biogas stations in Europe

In the last decade, the direction on the use of bio gas in the form of bio methane is actively developing. Biomethane is bio gas, brought to the quality of natural gas. However, for this purpose it is necessary to provide for the purification of bio gas from carbon dioxide, hydrogen sulfide and other impurities, which leads to an increase in the costs of its production. Installations for the production of bio gas with an increased content of methane have been called bio methane plants. In 2016, the number of bio methane plants in the EU countries is 367 units. The largest number of plants are in Germany (178 units), Sweden (59 units) and Great Britain (37) (Figure 2) [4].

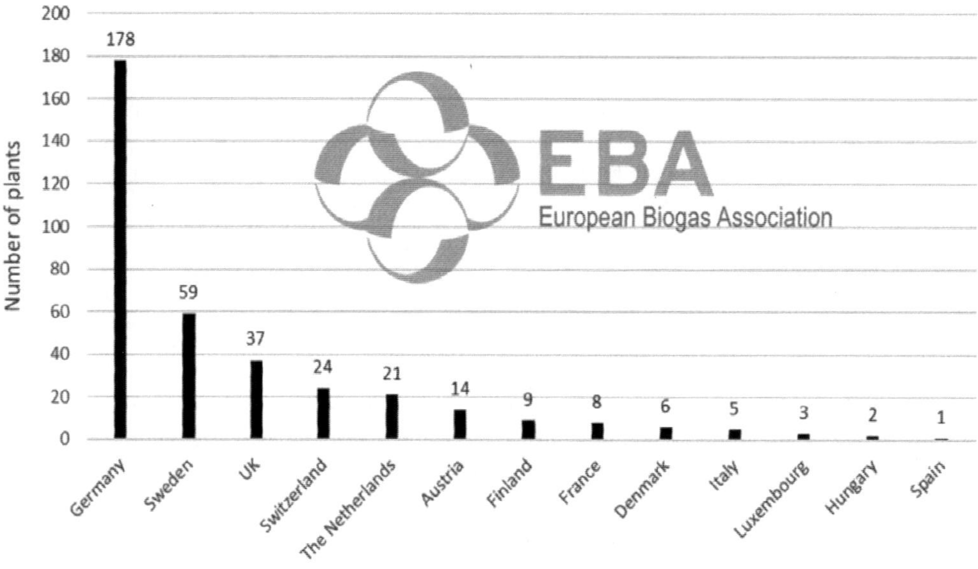


**Figure 2.** Number of biomethane plants in Europe

In Sweden, most bio methane plants supply bio methane to automobile gas-filling compressor stations (AGNKS) for further refueling with compressed bio methane.

It should be noted that in Europe the produced bio-methane is mainly used for supply to centralized gas supply systems (more than 200 plants) [4].

The receipt and use of bio methane is most promising in regions with a high level of development of the agricultural sector. This problem is most relevant for the Belgorod region, which produces about half of all meat products of the Central Federal District of the Russian Federation. In the Belgorod region, the annual amount of waste produced by the agro-industrial complex is more than 15 million tons. The total production potential of bio methane from waste generated is estimated at 465 million m3 / year, which is 14% of the annual volume of gas consumption in the Belgorod region.

The bio gas plants operating and being built on the territory of the Russian Federation use the produced bio gas in cogeneration plants for generating electric and thermal energy. When bio-gas is burned in cogeneration plants, it is previously cleaned of mechanical impurities, moisture and hydrogen sulfide. The main problem, restraining, growth of production of bio methane is the need to purify from carbon dioxide and, accordingly, the installation of expensive equipment.

In the Russian Federation, natural gas used in industry and the public utility sector is governed by state standard, which regulates the composition of the gas and the permissible content of non-combustible impurities (Table 1) [5].

**Table 1.** Physico-chemical parameters of gas combustible naturalindustrial and municipal purposes

| Indicatorname | Norm |
|---|---|
| Componentcomposition, molarfraction, % | Not standardized. The definition is mandatory |
| The lowest heat of combustion at standard conditions, MJ / m (kcal / m) | Not less than 31,80 (7600) |
| The range of the Wobbe (higher) number under standard conditions, MJ / m (kcal / m) | From 41,20 to 54,50 (from 9840 to 13020) |
| Mass concentration of hydrogen sulphide, g / m | No more 0,020 |
| Mass concentration of mercaptan sulfur, g / m | No more 0,036 |
| Molarfractionofoxygen,% | No more 0,050 |
| Molar fraction of carbon dioxide,% | No more 2,5 |
| Mass concentration of mechanical impurities, g / m | No more 0,001 |

Given these requirements for the supply of bio methane in the gas supply system, it is necessary to achieve a reduction in carbon dioxide content of less than 2.5 mole. %.

In Europe, the following cleaning methods are used to remove carbon dioxide ($CO_2$) [6]:

- adsorption under variable pressure;
- absorption by water under pressure;
- absorption by amine;
- absorption by other substances;
- membrane cleaning method;
- cryogenic method of cleaning.

The most widespread application was obtained by water absorption, adsorption under variable pressure and absorption by amines.

The Russian Federation has accumulated considerable practical experience in the field of natural gas purification from carbon dioxide. Equipment for the removal of $CO_2$ is widely used in gas fields in the extraction of natural gas and oil. However, taking into account the high content of $CO_2$ in the bio gas produced, reaching 20-30%, it is necessary to conduct a complex of studies aimed at finalizing the equipment to achieve the required degree of purification.

After purification, bio methane can be fed to the gas transmission network for distribution and supply to consumers. Depending on the volume of bio methane produced, it is necessary to determine the method and place of supply to centralized gas networks. To supply bio methane to the main gas pipelines, it is necessary to provide large volumes of gas with a high over pressure (3-5 MPa), which will require the installation of an additional compressor station and lead to an increase in capital expenditures. Taking into account that during purification of bio methane the pressure of the gas leaving the treatment facilities reaches 0.3 Mph, it is more expedient to supply bio methane to the distribution pipelines of medium and low pressure.

It can be concluded that the use of bio gas as a bio methane for centralized gas supply systems is a promising direction for the development of the gas industry. It has been established that the use of bio methane is most expedient in regions with a high level of development of the agricultural sector. Thus, the potential for bio methane production from agricultural waste in the Belgorod region is 465 million $m^3$ / year, which is 14% of the annual gas consumption in the region.

**Acknowledgments**
Work is performed within realization of the grant of the President of the Russian Federation SP - 1716.2015

**References**

[1] Development of the biofuel market in the world and in the Russian Federation 2012 *Russian Energy Agency* [Electronic resource]
[2] Kuschev L, Suslov D 2015 Theoretical description of the process of anaerobic fermentation in biogas plants *Bulletin of the Belgorod State Technological University V G Shukhov* vol 6 pp 227-230
[3] Biogas Barometer*A study carried out by EurObserv'ER* [Electronic resource]
[4] Klinkmuller L Studying the German experience in the production and use of biogas biomethane as alternative fuels*Mathematics of the NAK Naftogaz of UkraineKiev* [Electronic resource]
[5] The state standard 5542-2014 *Gases combustible natural industrial and municipal purposes Technical specifications*
[6] Geletukha G, Kucheruk P, Matveev Yu 2014 Prospects of production and use of biomethane in Ukraine *Analyticalnote BAU* vol 11

# COMPARISONOF USE EFFICIENCY OF DIFFERENT TYPES OF DENSIVE BALL LOAD DURING THE GRINDING OF PORTLANDCEMENT CLINKER

**V D Barbanyagre and A A Stronin**

Department of Cement and Composite Materials Technology, Belgorod State Technological University of V.G. Shukhov, Belgorod, 308012, RU

**Abstract**. The authors of this article carry out a comparative analysis of the effectiveness of using several types of dense ball packaging for cement production. This analysis is based on the study of the degree of influence on the kinetics of the clinker grinding process in the open cycle mill of different assortment of balls without the use of the milling intensifier and in the case of its use. The possibility of refusing the use of an intensifier in the case of rational selection of grinding bodies.

Clinker is characterized by a small amount of grinding, so the process of its grinding is associated with a large energy cost of 45 ± 5 kWh / t cl. [1]. As is known, the productivity of a cement mill is limited by the aggregation of the particles of the ground material, by adhering them to grinding bodies and armored, thereby reducing the efficiency of the grinding process in the grinding unit, which is accompanied by an increase in the specific consumption of electricity [2].

This circumstance makes it necessary to find ways to reduce the specific electricity consumption of cement mills and increase their productivity. In this paper, the application of some of them has been considered: the use of a dense spherical packing (PSU) [3] and an intensifier of grinding [4].

The aim of the work was to study the influence of different types of dense ball charge on the disperse characteristics of clinker, such as the specific surface and residue on the screen No. 008, and also to assess the degree of their effect on these parameters in the presence of a grinding intensifier, which was used by "Lithoplast 4I"PolyplastNovomoskovsk.

The investigations were carried out at the clinker of the Oskolcement plant. The clinker of the fraction 10-0 mm without addition of gypsum to it was fed to the GIPROCEMENT mill of 0.5 × 0.28 m. The weight of the feed material in all the experiments was 4 kg.

The material was crushed in the first chamber for 10 minutes, in the second for 15 minutes and 10 minutes. Through these intervals of clinker grinding time, samples were taken to establish the specific surface area and the total residue on a standard No. 008 sieve. The specific surface was determined by the air permeability method.

The particle size distribution of Portland cement clinker, milled in a mill for 35 min, was determined on a laser particle analyzer ANALYSETTE 22 NanoTec plus from FRITSCH.

The whole process of grinding passed in one chamber of the mill. In the first chamber, a constant range of balls was always loaded: Ø74 / Ø54 with a 2: 1 ratio of balls ($\varphi = 0.2$, mass of grinding bodies 55 kg), then after 10 min of clinker grinding from the mill, this ball charge was extracted, and another assortment of balls was loaded, corresponding to the second chamber: Ø60 / Ø17 = 1: 2 or Ø40 / Ø17 = 2: 1.

In the case of combined loading, the material was frayed in only one chamber for 35 minutes, with sampling at the above time intervals.

The studies were started with a combined loading: Ø74 / Ø54 / Ø40 / Ø17 = 1.3: 0.4: 2.2: 1.0; $\varphi = 0.2$, the mass of the balls is 55 kg, the power capacity is 13.75. After 35 minutes of grinding, the specific surface area of the clinker is 329.4 m2 / kg, the residue on the No. 008 screen is 7.1%, the content of the fraction is 5-30 μm = 68.9%.

In the case of using a tight ball packing with diameters of balls Ø60 / Ø17, a mass of a grinding load of 55 kg, $\varphi = 0.2$ and an energy capacity of 13.75 after the same time interval, the indicated parameters have the following values: Ssp = 345.49 m2 / kg, $R_{008}$ = 1.34%, 5-30 µm = 74.15%.

The use of a treadmill Ø40 / Ø17, having a mass of grinding bodies of 80 kg, a load factor of 0.3 and a power-bearing capacity of 20, leads to the following results: $S_{sp}$= 411.28 m2 / kg, $R_{008}$ = 3.94%, 5-30 µm = 69 , 59%.

At this stage of the research from the data obtained (Table 1), it seems that the rational assortment of grinding media is a tight ball packing Ø60 / Ø17, because it makes it possible to obtain a material with a residue on a standard sieve No. 008 5 times less than when using a combined load and 3 times less, compared to a tight ball load Ø40 / Ø17. By the content of the fraction of 5-30 µm, all three samples have approximately the same value.

**Table 1.**Effect of assortment of milling bodies on disperse characteristics of Portland cement clinker

| Assortmentofgrindingload | The ratio of fr. 5-30 microns to fr. 0-5 microns | Specificsurface, m2 / kg | Residueonscreen No. 008,% | Contentsfr. 5-30 µm,% |
|---|---|---|---|---|
| Combineddownload | 4,7 | 329,4 | 7,1 | 68,9 |
| PSN Ø60/Ø17 | 5,1 | 345,5 | 1,3 | 74,1 |
| PSN Ø40/Ø17 | 4,7 | 411,3 | 3,9 | 69,6 |

Then, the previous experiments were repeated, but with the addition of the grinding aid (0.1% by weight) to the second chamber, it was not first introduced. With the combined loading, the additive was introduced into the mill after 10 minutes of clinker milling, in order to enable us to further compare the results obtained. The data of the carried out researches are resulted in tab. 2.

**Table 2.**The cumulative effect of the assortment of grinding bodies and the grinding intensifier on the disperse characteristics of Portland cement clinker

| Assortmentofgrindingload | The ratio of fr. 5-30 microns to fr. 0-5 microns | Specificsurface, m2 / kg | Residueonscreen No. 008,% | Contentsfr. 5-30 µm,% |
|---|---|---|---|---|
| Combineddownload | 6,6 | 377,3 | 1,4 | 87,9 |
| PSN Ø60/Ø17 | 5,9 | 358,8 | 0,2 | 83,3 |
| PSN Ø40/Ø17 | 5,4 | 394,1 | 2,4 | 85,5 |

The addition of the additive to the combined charge leads to a decrease in R008 of approximately 80%, an increase in the specific surface area and the content of fr. 5-30 µm approximately by 14.5% and 28%, respectively.

The addition of the additive to the grinding chamber with a trough Ø60 / Ø17 is only accompanied by a 85% reduction in the total residue on the No. 008 screen. Change in the specific surface area and content of fr. 5-30 µm is not significant.

The same result is given by the replacement of the camera 60 Ø17 / Ø17 on the camera Ø40 / Ø17: only a decrease of R008 by 39% is observed. The remaining parameters hardly change.

From this it can be seen that the input to the grinding mill intensifier reduces the total residue on the No. 008 sieves by 80%, in case of using only a combined load in the second chamber or a Ø60 / Ø17 camera.

All the results obtained in the course of this work are given in general form in Fig. 1 and 2.

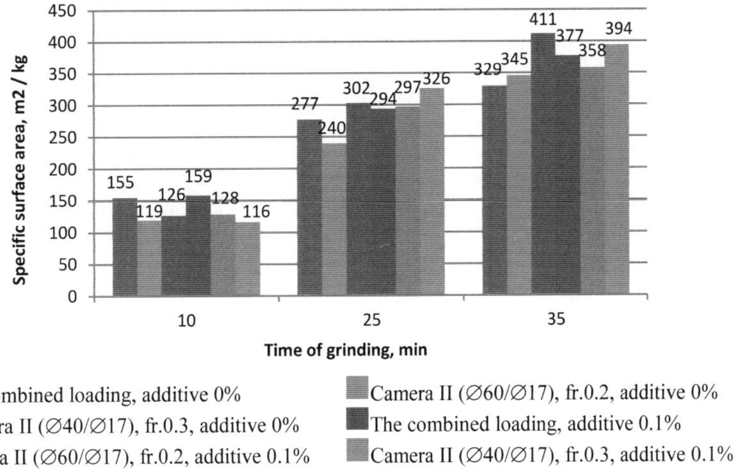


**Figure 1.** Dependence of the specific surface area on the grinding time

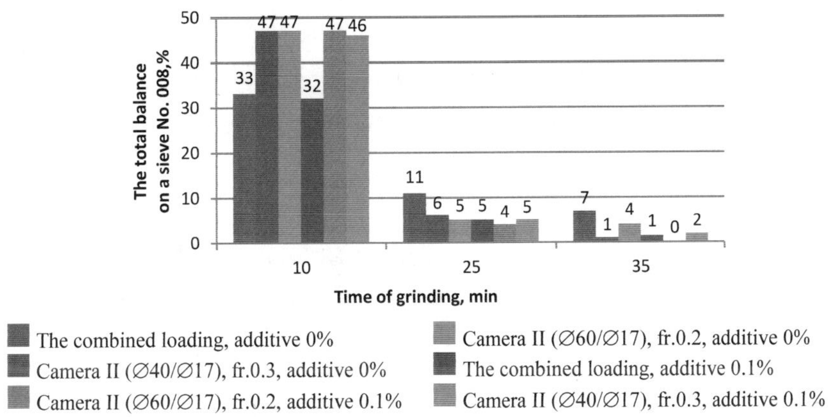


**Figure2**. Change in the value of the total residue in the grinding process

According to the results of the research, it is established that the camera Ø40 / Ø17 is inferior to the camera 60 / Ø17 in the residue on the control screen No. 008, in the case of clinker grinding in the mill for more than 25 minutes, at an early stage they have the same efficiency. In the production of ordinary cement, the use of both types of dense ball packing is expedient in the absence of a grinding intensifier.

If we accept the value of the specific surface in the range of 300-320 $m^2$ / kg as the criterion for the readiness of the product, we can make a comparative assessment of the effectiveness of using different types of assortment of balls. The mill's productivity in this case will depend on the time it takes to achieve the specified fineness of grinding.

So, in case of using a tight ball packing with diameters of balls Ø40 / Ø17, this will take 25 minutes, in all other cases - about 30 minutes. This leads to the important conclusion that the use of the grinding intensifier does not always lead to an increase in the productivity, and in the case of a competent selection of the range of grinding load, it is no longer necessary to use it for the production of ordinary cement.

The carried out researches prove expediency in continuation of researches on studying of influence of an assortment of grinding bodies on kinetics of process of grinding clinker in a ball mill. The paper shows that rational selection of grinding bodies allows to obtain cement with high dispersive characteristics, without the use of grinding intensifiers, which ensures a reduction in the cost of the cement produced by the enterprise, as well as an increase in the mill's productivity and a decrease in the specific consumption of electric energy.

**Acknowledgments**
The article was prepared within development program of the Flagship Regional University on the basis of Belgorod State Technological University named after V.G. Shukhov.

IOP Conf. Series: Journal of Physics: Conf. Series **1066** (2018) 012005 doi:10.1088/1742-6596/1066/1/012005

**References**

[1] Klassen V K 2012 Technology and optimization of cement production(Belgorod: BSTU Publishing House V.G. Shukhova) p 308
[2] Khodakov G S 1972 Fine grinding of building materials(Moscow: Publishing House of Literature on Construction) p 239
[3] Matveev A F, Smal D V, Moskvichev D S , Barbanagre V D 2015 Insensification of fine grinding of materials in ball tube mills(Publishing House of BSTU) pp 20-27
[4] Karibaev K K 1980 Surface-active substances in the production of soldering materials (Alma-Ata: Science Publishing House) p 336

IOP Conf. Series: Journal of Physics: Conf. Series **1066** (2018) 012006 doi:10.1088/1742-6596/1066/1/012006

# CONSTRUCTIVE AND TECHNOLOGICAL FEATURES OF UNITS FOR MILLING POLYMER MATERIAL

**V S Sevostyanov , P Y Goryagin, M V Sevostyanov and R A Ermilov**

Department of Technological Complexes of Machines and Mechanisms, Belgorod State Technological University of V.G. Shukhov, Belgorod, 308012, RU

**Abstract**. This article presents specific physical and mechanical properties of techno genic polymer materials. The features of their processing are revealed. The ways of constructive and technological perfection of grinders are proposed, namely, the creation of units based on needle-milling working parts that ensure selective complex influence on the processed materials with the cantilevered ends of their rod elements and the creation of energy-efficient mechanisms of destruction.

At present, various methods for the utilization of techno genic polymer materials (TPM) are known, among which we can mention the processing of biodegradable polymers, thermal processing, chemical processing and recycling of polymeric materials. There are sufficient incentives for recycling of utilized polymer waste such the environmental aspect, consumer demand, legal requirements and low cost [1].

Secondary recycling of used polymers is an important task. Although the content of TPM products in wastes is relatively small (about 7-8% by weight),the problem of utilization of these wastes (about 18% 20% by volume) is significant because of a low specific gravity [2].

Due to their high resistance to environmental influences, these materials remain in natural conditions for a long time. On the basis of laboratory experiments, it was possible to establish thatit will take from one hundred to five hundred years for a full decomposition of a plastic bottle under the influence of ultraviolet and temperature drops. Some environmentalists also indicate a period of up to 1000 years [3]. At the same time, production from polymer materials is more than three hundred million tons every year and continues to grow. In this regard, it can be argued that TPM is one of the most dangerous pollutants of the environment.

Polymeric materials have a wide range of physical and mechanical characteristics, which strongly depend on their structure. In addition,various external factors in the process of processing can influence on the mechanical properties of the substancea lot [1-3]. These include: thermal impact; frequency, duration or rate of loading; type of stress state; pressure; nature of the environment, etc. A feature of the physical and mechanical properties of polymeric materials is their relatively high strength at very low rigidity (in comparison with metals) [4].

In this regard, the aim of this paper is to analyze the structural and technological characteristics of units for grinding polymeric materials and to develop energy-efficient equipment for the processing of TPM.

Polymers are usually divided into solid, the modulus of elasticity of which is equal to E = 1..10 GPa (fibers, films, plastics), and soft highly elastic substances, the modulus of elasticity is E = 1..10 MPa (rubber). The regularities and the mechanism of destruction of these and other are different [5]. In some cases, in order to increase the productivity of the grinders, it is necessary to pre-prepare the polymer materials before the crushing stage beforehand. The type of the preparatory technological operation depends on the physical and mechanical characteristics of the waste, on their quantity, geometric dimensions, etc. For example, materials having a small bulk density must be compacted before grinding beforehand [6].

When developing energy-efficient equipment for grinding TPM, it is necessary to take into account their specific physical and mechanical properties. A new direction of constructive and technological perfection of grinders includes the creation of multifunctional technological units based on needle-milling working parts that ensure selective complex effect on the processed materials by the cantilevered ends of their rod elements and the creation of energy-efficient mechanisms of destruction [7-8]. Of the whole variety of known sets of rod that form wire brush of various designs disc, cylindrical, flat, finger and etc. wire brushes canbedistinguished asthe most common ones [9].

The task of intensifying the process of grinding and the expansion of technological possibilities is materialized in the device for grinding of techno genic materials with wire brush working parts[10] (Fig.1, 2).

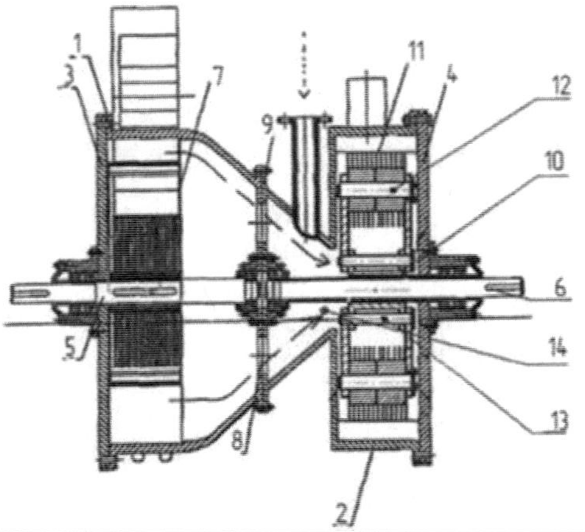


**Figure 1**. Plant for grinding of techno genic materials: 1, 2 - cylindrical-camera; 3, 4 - the cover; 5.6 - shaft; 7 - partition of the first chamber; 8 –transitional aperture; 9 - inter-chamber partition; 10 - the hub; 11 - the brush;
12 - an axis; 13 - cotter pin; 14 - inlet openings of the second chamber

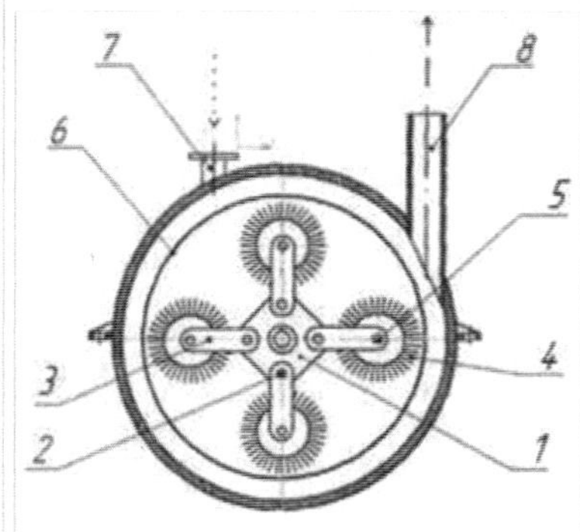


**Figure 2**. The chamber of a fine grinding: 1 – the hub; 2, 5 - an axis; 3 - the earring;
4 - brush; 6 - a drum; 7 - branch pipe of additives; 8 - discharge outlet

The plant comprises a body consisting of two horizontally placed chambers: coarse grinding 1 and fine grinding 2. According to the invention, the wire brush working parts of the grinding chamber are made of a set of rod elements assembled into bags and rigidly fixed by one end on the holder. Such a rotor has a developed working surface, which provides efficient grinding of the material, and expands the technological capabilities of the unit.

Thus, it can be concluded that the development of devices with wire brush working bodies is an innovative direction in the structural and technological perfection of grinders, taking into account specific physical and mechanical properties of TPM. The use of wire brush tools increases the energy efficiency of grinders by increasing the number of single contacts of the working element and material. Besides, shocking, shearing, cutting and tearing stresses in contact zones occur simultaneously, which not only intensifies the process of grinding polymeric materials, but also ensures the one-dimensionality of the product obtained.

IOP Conf. Series: Journal of Physics: Conf. Series **1066** (2018) 012006 doi:10.1088/1742-6596/1066/1/012006

**Acknowledgments**
The article was prepared within development program of the Flagship Regional University on the basis of Belgorod State Technological University named after V.G. Shukhov.

**References**

[1] Gogol E V, Mingazetdinov I Kh, Gumerovaia G I2013 Analysis of existing methods of waste treatment and recycling polymers *Vestnik of the Kazan Technological University* **10** pp 163-168
[2] SverguzovaS V 2008 Industrial ecology as an indispensable condition for stable development (Belgorod: Publishing House of the Belarusian State Technical University) p155
[3] SevostyanovV S,NosovOA, VardanyanG R, Berezhnoy O L 2013 Processing of used polyethylene packaging Ecology: Education, Science, Industry and Health: Sat. doc. *V International Scientific and Practical Conf.* (Belgorod: BSTU Publishing House), pp 152-156
[4] Bystrov GA, Galperin VM, Titov B P 1982 Deactivation and disposal waste in the plastics industry ( L .: Chemistry) p 264
[5] SevostyanovV S, NosovOA, BerezhnoyO L, Vardanyan G R 2015 Technique and technology of grinding secondary polymeric raw materials *Energy and Resource-Saving Environmentally Clear Chemical Processes of environmental protection:* Sat. doc. international scientific and practical conference (Belgorod: Publishing House of BSTU) pp 50-54
[6] Sevostyanov V S, MikhailichenkoS A, Sevostyanov M V 2007 Energy-saving technological complexes and units for utilization of techno genic materials *Vestnik of the Belgorod State Technological University named after V.G. Shukhov*vol 1 pp 85-90
[7] Sevostyanov VS, Sivachenko TL 2016 Wire brush grinders, their technological possibilities and ways of development Vestnik of the Belorussian-Russian University**1**pp69-78
[8] Sevostyanov VS, Sivachenko TL, Mikhaylichenko S A 2015 Technological devices with wire brush parts organs for complex processing of composite materials *Vestnik of the Belgorod State University Technological University V.G. Shukhov*vol 2 pp 50-57
[9] SevostyanovV S, UralV I, SevastyanovM V, Nosov OA 2015 Technological complexes and equipment for processing and utilization of technogenic materials*Belgorod: Publishing house BSTU V.G. Shukhova* p 321
[10] Glagolev SN, Sevostyanov VS, Sevostyanov MV Pat. 2540549 Ros. Federation, IPC B02C 18/00*Plant for grinding fibrous materials and etc.*; applicant and patent holder of BSTU V.G. Shukhov No. 2013141378/13; claimed. 09/09/2013; (publ. 10.02.2015, **4**) p 11

IOP Conf. Series: Journal of Physics: Conf. Series **1066** (2018) 012007 doi:10.1088/1742-6596/1066/1/012007

# ECONOMIC FEASIBILITY OF VENTILATION CLEANING EMISSIONS IN UNDERGROUND PARKINGS

**T N Il'ina and V V Isaeva**

Department of Heat and Gas Supply and Ventilation, Belgorod State TechnologicalUniversity of V.G. Shukhov, Belgorod, 308012, RU

**Abstract**. The expediency of purification of exhaust gases from vehicles in underground parking lots is shown. Methods of trapping harmful substances in the ventilation air of parking lots are considered. Equations are presented for calculating the index of environmental and economic damage to the environment from vehicles. The calculation of ecological and economic damage is given in the example of underground parking in the city of Anapa.

Every year the ecological situation in our country is aggravated. One of the main air pollutants is automobile exhaust. The most polluted car exhaust occurs during engine start-up and its warm-up, which means in car parks and parking lots. Manufacturers of cars year after year are trying to improve the design of engines and exhaust systems, but the problem of protecting the environment from harmful emissions is archival.

Due to the fact that the car most pollutes the environment at the time of starting and warming up the engine, the ground open car parks are not able to detain and purify air from exhaust gases, and the construction of ecological parking in our country is not as developed as in European countries. Thus, the most feasible is the construction of covered ground and underground parking lots [1, 2], including devices for trapping harmful gases.

Objective:

- assess the degree of air pollution when starting and warming up the engine;
- consider methods of air purification from harmful emissions;
- choose a way of quantitative assessment of environmental and economic damage to the environment from vehicles.

As already mentioned above, the most acute environmental situation occurs in the parking lot and car parking areas.

Such nonstationary regimes, including the heating of a cold engine, take no more than 3-5 minutes in a warm season and from 15-30 minutes to 1-2 hours during the cold season. At the same time, the operation of the engine in such modes is accompanied by a significantly greater emission of harmful substances with the exhaust gases (up to 8-10 times) than in stationary operating modes. The "cold" car consumes 27% more fuel than the "hot" car, and at the same time it selects more CO by 86%, CH - 40%, NOx - by 12% [3].

The composition of exhaust gases includes about 200-300 chemical compounds and substances (Table 1). Of these, only four in small concentrations do not harm the body (oxygen, nitrogen, carbon dioxide and water).

There are several methods for purifying air from vapor-gaseous pollutants, by the type of physico-chemical processes they are used, they are divided into five main groups [4]:

1) rinsing out solvent emissions of impurities (absorption);
2) washing solutions with solutions of chemicals that bind impurities chemically (chemisorption);
3) absorption of gaseous impurities by solid bodies with an ultra-microporous structure (adsorption);
4) thermal neutralization of waste gases;
5) absorption of impurities by using catalytic conversion.

For the purpose of complete purification of gas emissions, it is advisable to combine methods in which the combination of coarse, medium and fine purification of gases and vapors is optimal for each particular case. In the first stages, when the content of a toxic impurity is large, absorption methods are more suitable, and for adsorption, adsorption or catalytic methods [5, 6]. The catalytic method differs from the thermal one, first, by the lower temperature of the oxidation process, 300-400 °C, and secondly by its high rate of flow, a fraction of a second, which makes it possible to significantly reduce the dimensions of the reactor. The catalysts can be platinum metals, oxides of copper, manganese, etc.

One of the most promising ways of protecting the environment and efficient energy use is the recovery of the heat of ventilation air in underground parking lots [7-10].

**Table 1.Composition of exhaust gases**

| Exhaustgascomponents | Contentbyvolume,% PetrolDiesel Contentbyvolume,% | Engine | Engine |
|---|---|---|---|
| Nitrogen | 74,0 - 77,0 | 76,0 - 78,0 | No |
| Oxygen | 0,3 - 8,0 | 2,0 - 18,0 | No |
| Watervapor | 3,0 - 5,5 | 0,5 - 4,0 | No |
| Carbondioxide | 5,0 - 12,0 | 1,0 - 10,0 | No |
| Carbonmonoxide | 0,1 - 10,0 | 0,01 - 5,0 | Yes |
| Nitricoxide | 0 – 0,8 | 0,0002 – 0,5 | Yes |
| Hydrocarbons | 0,2 - 3,0 | 0,009 - 0,5 | Yes |
| Aldehydes | 0 - 0,2 | 0,001 - 0,009 | Yes |
| SulfurOxide | 0 - 0,002 | 0 - 0,03 | Yes |
| Carbonblack, g / m3 | 0 - 0,04 | 0,01 - 1,1 | Yes |
| Benzpyrene -3.4, g / m3 | 10 - 20×10$^{-6}$ | 10×10$^{-6}$ | Yes |

In connection with the arrival of a large number of polluting substances in the atmosphere, there is a need to quantify the impact of vehicles on the environment. Among the economic indicators that characterize the environmental problems of interaction between vehicles and the environment, the indicator of environmental and economic damage has been studied quite well. This indicator consists in determining actual and possible material and financial losses from changes in quantitative and qualitative parameters of the environment as a whole, as well as its individual natural components (water and land resources, atmospheric air, etc.) [11-13].

In accordance with the Temporary Method for Determining the Prevention of Environmental Damage, an estimate of the magnitude of environmental damage from air pollution is a [14]:

$$y^a = y^a_{sp} \cdot M^a_p \cdot K^a_e \quad (1)$$

where$y^a$ – is the value of the economic estimate of pollutant emissions into the atmospheric air, rub; $y^a_{sp}$ – specific environmental and economic damage from the emission (discharge) of one conventional ton of pollution, rub./ conditional tons; $M^a_p$ – reduced mass emission of pollutants, conditional tons/ year; $K^a_e$ – coefficient that takes into account environmental factors for the territory under consideration (for the North Caucasus Economic Region is 1.92).

The resulted mass of emissions is defined by the formula:

$$M^a_p = \sum_{i=1}^{N} m_i \cdot A_i \quad (2)$$

where$m_i$ – the actual mass of the release of the i-$^{th}$ type of pollutant per year, ton/year; $A_i$ – relative hazard index of the i-$^{th}$ type of pollutants, conditional tons/ ton [14].

Specific environmental and economic damage from the release of one conventional ton of pollutants is calculated by the formula[15]:

$$У^a_{sp} = У^a_{sp1998}(1 + r)^m \quad (3)$$

where$У^a_{sp.1998}$– environmental and economic damage from the release of one conventional ton of pollutants into the air for 1998 year, component for the North Caucasus region 53,2 rub./conditional tons; $r$ – rate of inflation; $m$ – number of years.

Thus, this method provides an assessment of the environmental and economic damage from the emission of pollutants into the atmosphere in the current period. Based on the example of the city of Anapa in the North Caucasus Economic Region, the environmental and economic indicator of environmental damage from underground parking for 106 parking places was 1 432 447 rubles per year. There are also other, more extensive, methods for determining environmental damage to the environment, the determining factor of which is, for example, the health and working capacity of the population [3].

In conclusion it should be said that the example of underground parking in the city of Anapa was used to calculate the environmental and economic indicator of damage to the air from car exhausts, which amounted to 1,432,447 rubles per year. To reduce environmental and economic damage, it is necessary to provide for the purification of vehicle exhaust fumes. At present, the most universal method is post-gas purification of gases by a sorption-catalytic method.

**Acknowledgments**
The article was prepared within development program of the Flagship Regional University on the basis of Belgorod State Technological University named after V.G. Shukhov.

**References**

[1] Kondrashin B 2016 Ecological parking as an innovative form of nature use Scientific community of students of the XXI century. Natural Sciences: Sat. Art. by mat. XXXIX international student scientific-practical Conference vol 3 pp 80-87

[2] Isaeva V, Litvinov I 2017 Analysis of ventilation systems of underground parkingInternational scientific and technical conference of young scientists BSTU V G Shukhov

[3] Levda N, Postnikov V 2013 Assessment of environmental damage to the population and the economy of the region from pollution of atmospheric air Economics of nature management vol 25 pp 37-45

[4] Vetoshkin A G2006 Processes and devices for gas cleaning: a training manual (Penza: Izd-vo PGU) p 201

[5] Batenkov V A 2002 Biosphere protection: educational-methodical manualBar-naul: Publishing house Alt. University p 193

[6] Trubaev P A, Tarasyuk P N 2017 Evaluation of energy-saving projects for generation of heat and heat supply by prime cost forecasting methodInternational *J of Energy Economics and Policy* **7** pp 201-208

[7] Nailova VN, Toporkova A A 2017 Methods of cleaning air from harmful substances from the emissions of closed auto-parkings *J Electronic scientific journal* **4-2** pp 375-378

[8] Ilyina T N, Belmaz D N 2014 Analysis and methods of utilization of secondary energy resources of the oil refinery *Bulletin of the Belgorod State Technological University named after. V.G. Shukhov* **3** pp 189-193

[9] Ilyina T N, Belmaz D N 2013 Ways of using waste heat from primary oil refineries *Bulletin of the Belgorod State Technological University named after. V.G. Shukhova* **5** pp 198-202

[10] Ilyina T N 2010 Recycling of secondary heat in production plants of bakery enterprises Bulletin *of the Belgorod State Technological University named after. V.G. Shukhov* **3** pp 147-150

[11] Nailova V N, Logachev I N, Baromychenko A A, Nebyltsova I V 2016 Assimilation of exhaust gases of a car in a closed car park Molodezh and scientific and technical progress: Collected papers of the IX International Scientific and Technical Conference of students, graduate students and young scientists (Stary Oskol: OOO "*Assistant Plus*") pp 195-198

[12] Tsyplakova E G, Yankevich Yu G 2016 Ecological and economic assessment of ensuring environmental safety of motor vehicles in the parking area *Economics of the New World* **3** pp 87-105

[13] Postnikov V P 2014 Analysis of air pollution: national and regional aspects *Bulletin of the VolSU. Series 3: The Economy. Ecology* **1** pp 117-124

[14] Sokolov L I, Kozlova A G 2001 Ecological and economic efficiency of enterprises (Vogloga: VSTU) 60 p

[15] Temporary methodology for determining preventable environmental damage: Chairman of the State Committee of the Russian Federation for Environmental Protection 09.03.1999.

IOP Conf. Series: Journal of Physics: Conf. Series **1066** (2018) 012008 doi:10.1088/1742-6596/1066/1/012008

# ECONOMICS AND ECOLOGY SYMBIOSIS: TRANSITION TO PRINCIPLES OF BEST AVAILABLE TECHNIQUES

**E G Gasho[1] and I A Gasho [2]**

[1] Department of Energy Conservation Methodology of National Research University Moscow Power Engineering Institute, Moscow, 111250, RU
[2] Department of Economic of Belgorod University of Cooperation, Economics and Law, Belgorod, 308023, RU

**Abstract**. The paper presents the successful introduction of BAT depends greatly on the balance of legislation in whole, which takes into account different aspects of state regulation and promotion of energy efficiency in the industrial complex as well as close link of technological, energetic and ecological indicators of efficiency and non-wastefulness.

In accordance with the Federal Law on 21 July 2014 No 219-FZ, large enterprises of the key sectors of industry with serious negative impact on environment and high natural resource consumption (belonging to Category I) are obliged to meet the requirements of the best available techniques (BAT) implementation in Russia and therefore to adopt integrated ecological solutions (IES) within the coming years.

BAT have boasted great potential for the development since the adoption of the Federal Law No 219-FZ in 2014 leading to the amendments in the Law on environmental protection and introducing the scheme classifying enterprises into different categories according to their environmental influence starting with the ones having the greatest impact. It was done to promote their gradual shift to BAT. Therefore, the frames of the scheme, which should be followed up with "vertical" (on different sectors) and "horizontal" (of general methodology) informational and technical guides, have been set. "Horizontal" guides and, in particular, the guide of the best available techniques have no industry specifics and are applicable to any sectors and industries considering certain technological aspects attributable to definite business spheres.

However, if we look at the new ecological legislation and analyze enterprises from its point of view, then, we see that the integrated proportion of the enterprises related to Group I in general consumption of fuel and energy resources (FER) and environmental pollution is considerable (Table 1).

**Table 1.**Integrated characteristics of Group I enterprises (in accordance with 219-FZ)

| Input | Entry |
|---|---|
| 17 % occupied | |
| 68,7 % of water | 68,7 % of sewages |
| 18 % of lands | 79,8 % wastes |
| 52 % of fuel and energy resources | 75 % emissions to the atmosphere |

Data processing of nearly 4500 Russian enterprises[1] has helped classify them due to the level of their technological and energy efficiency. The classification of enterprises on their energy and resource efficiency allows to identify the size of major groups, their sectoral belonging and to offer certain proper measures to move to other groups with more favorable indicators of energy and resource efficiency and environmental safety. The leading groups include about 20% of companies from various sectors (see Figure 1).

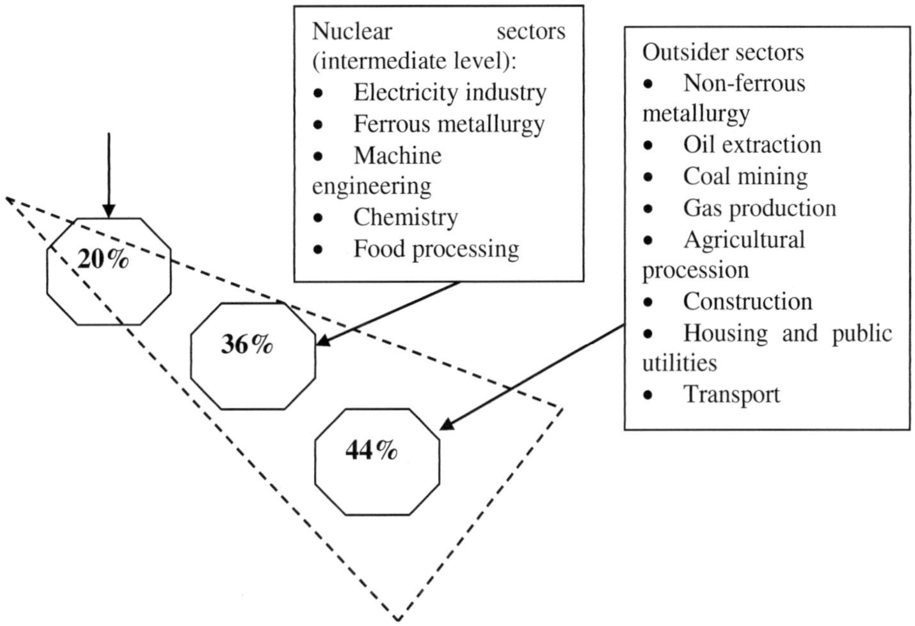


**Figure 1**. Distribution of industries on their energy and resource efficiency

Energy and resource efficiency indicators of almost 20% enterprises are above average level, about 36% are at the average level and 44% are below the average one. Therefore, the key target of the state authorities implies to relocate outsiders to the group of "intermediates" and, hence, to increase the group of "leaders".

According to the expert and analytical surveys, while buying equipment, almost 30% enterprises consider the uncertainty in its efficiency to be the core hurdle. With this regard, the availability of such guides might be particularly claimed as reference documents.

Environmental regulation (the reduction of negative impact of enterprises on the environment) is closely connected to energy efficiency increase. A significant part of positive ecological results is achieved by implementing measures and programs of energy intensity reduction. The society of professionals place a lot of hope on the guides of the best available techniques in terms of both comprehensiveness and balance level of the state attitude to BAT and from the point of view of direct involvement of all the entire set of actors.

For instance, without any certain frames of the system under the research, we face the first important methodological issue that enquires what indicators should be taken as the measure of resource, energetic and ecological efficiency for the technologies of energy and technological machines and assemblies, in particular, or workshops and enterprises in whole. Since a significant number of high tem-

[1]The assessment based on the data processing of 4200 enterprises http://interfax-era.ru/reitingi-predpriyatii/fundamentalnaya-effektivnost/sredneotraslevye-znacheniya

IOP Conf. Series: Journal of Physics: Conf. Series **1066** (2018) 012008 doi:10.1088/1742-6596/1066/1/012008

perature heat systems is equipped with recycling appliances, we can use the indicators therewith this recycling process defining their energy and ecological efficiency. (Table 2). In some cases, it can lead to the considerable reduction of indicators of physical energy capacity and specific atmospheric emissions per unit of produced goods.

**Table 2.**The list of energy and resource efficiency indicators

| Efficiency indicators | Ways of determining the indicators |
|---|---|
| Visible fuel consumption, efficiency ratio of assembly (system) | $B_v = B_e / G_{con}$, $\eta = Q_{eff} / Q_{com}$ |
| Specific consumption with regeneration and recycling of secondary energy resources | $B^*\ sp = (B_v - B_{reg})/G_{con}$ |
| Specific wastes and emissions per unit of produced goods (fuel) | $w = W_{com} / G_{con}$ $w = W_{com} / B_{sp}$ |
| Specific wastes and emissions per unit of fuel with regeneration | $w^* = W_{com} / B^*_{sp}$ |
| Specific wastes and emissions per unit of produced goods with secondary products (steam, electricity, cold) | $w^* = W_{com}/(G_{con} + D_{rec})$ |

Here, B is for fuel consumption, G is for assemblies' efficiency, W is for emissions, $\eta$ – efficiency ratio, Q is for heat production, * is for values of indicators with secondary energy recycling.
In the process of developing guides, some important issues can occur, that is, for instance, what figure of the indicators in use can be set as the extreme limit of "the best" techniques. In turn, this may affect the way in which enterprises move towards BAT, if this transition is complex and expensive and what proportion of enterprises is considered inefficient by the moment the campaign starts.
The new era of environmental regulation in the Russian Federation set by the amendments in the Federal Law on 21 July 2014 No 219-FZ stipulates the close link among technological, energetic and ecological indicators of functioning of energy-intensive heat technological assemblies and aggregates at almost all industrial sectors.

In 2015 and 2016, 26 informational and technical guides (ITG) were issued. They included information related to pulp and paper industry, sectors of copper, aluminum, mineral fertilizers, cement, lime, ceramic pieces and glass production. Moreover, the guide on thermal neutralization of waste products (waste burning) was published either. It was the first time when the guides, which have no international analogues, were developed, namely, "Wastewater treatment when making products and rendering services at large enterprises" and "Wastewater treatment with the use of water supply and sanitation systems of settlements and urban districts". In 2017, the development of guides was completed and 51 guides on BAT were approved.

Importantly, the successful introduction of BAT depends greatly on the balance of legislation in whole, which takes into account different aspects of state regulation and promotion of energy efficiency in the industrial complex as well as close link of technological, energetic and ecological indicators of efficiency and non-wastefulness.

**References:**

[1] Begak M 2010 The best available techniques and integrated ecological solutions: prospects of application in Russia (Moscow: YurInfoR-Press)

[2] Reference document on the best available techniques of energy efficiency provision 2012 (Moscow: EcoLine)
[3] Lisienko V G 2014 The best available techniques of energy efficiency provision at energy-intensive enterprises Training and reference book (Yekaterinburg)

IOP Conf. Series: Journal of Physics: Conf. Series **1066** (2018) 012009 doi:10.1088/1742-6596/1066/1/012009

# FEATURES OF DETERMINING CRITERIA OF SIMILARITY FOR FREQUENCY OF ELECTRIC DRIVES

**M A Averbukh and S V Khvorostenko**

Department of Power Engineering and Automation, Belgorod State Technological University of V.G. Shukhov, Belgorod, 308012, RU

**Abstract**. The article describes the application of the similarity theory for a frequency-driven electric drive that allows the results of an exploratory model to be transferred to an object of similarity. At large industrial enterprises a large number of such electric drives of various designations are used, the experimental study of which is impossible to assess the impact on the technological process or the research is difficult due to the fulfilment of the technological task. For this it is necessary, on the basis of the π-theorem, to compile similarity criteria for power drives of different frequency, which are used in research and applied in an industrial enterprise.

Frequency drive is a set of electromechanical devices, including a frequency converter and an asynchronous motor [1]. The use of such devices provides a reduction in power consumption, depending on the operating mode up to 25%. Along with the optimization and improvement of the quality of the process, such electrical appliances are the cause of higher harmonics of current and voltage of the generators being fed into the power system [2]. To determine the degree of influence of non-sinusoidal mode on the power supply system for such electric receivers, it is necessary to conduct experimental studies on electromagnetic compatibility [3].

The complexity of such an experiment lies in the impossibility of investigating each frequency drive and the permissible change in its operating mode. In order to carry out the studies with different operating modes of the applied electric drives, it is necessary to develop criteria for the similarity of complex systems for an electric drive of lower power that is in accordance with an industrial electric drive. To this end, it is necessary to develop the criteria for the similarity of a frequency electric drive for the possibility of transferring the received characteristics from a research object to an industrial object.

The aim of the work is to develop similarity criteria and scale factors of different in frequency power electro-drives.

The similarity theory and the criteria developed with its help make it possible to establish the presence of similarity of complex systems or to develop methods for obtaining it in the form of scale ratios, complex functional dependencies of groups of parameters of the objects being compared. Similarity criteria are the establishment of the correspondence between the compared objects and the processes occurring in them, in which the functions of the objects are known, and the mathematical descriptions allow their transformation to the identical form [4].

For the construction of π-criteria, according to the first supplementary provision on the similarity of complex systems, to determine the likelihood of a frequency drive, we apply the method of structural modeling, on the basis of which it is possible to single out individual subsystems: an autonomous voltage inverter (AIN), an uncontrolled rectifier HB), LC-filter (PZPT) and the asynchronous motor (AD). Let us write down the generalized system of scalar equations of each element of a two-link frequency converter in a system of moving coordinates [u; υ], which are fundamental for the original object. Such a system of coordinate axes at $\omega_k = 0$ is used to describe processes in electric drives in which the output voltage of a semiconductor frequency converter is not sinusoidal, also the current in phase coincides with the u axis, and always corresponds to the real phase current [5].

The system of equations (3-6), where each element of the subsystem of a record with a prime, will refer to the similarity model. On the basis of an additional provision on the similarity of complex systems consisting of sub-systems and using the Fourier rule, we define π-criteria, using the reduction of equations to the dimensionless form with the help of integral analogues [6]. In this case, the compared processes in the original object and the similarity object are similar, therefore, between their similar parameters there must exist proportionality relations, an example for the third equation (5):

$$i_c = m_{i_c} \cdot i_c^{'}; \quad i_в = m_{i_в} \cdot i_в^{'}; \quad i_u = m_{i_u} \cdot i_u^{'}, \qquad (1, 2)$$

where$m_i$ – scale factors.

1.1. Equations of the AIN

$$f_u = \frac{\overset{*}{u_u}}{2 \cdot U_0}, f_\upsilon = \frac{\overset{*}{u_\upsilon}}{2 \cdot U_0},$$
$$u_{uu} = u_u \cdot f_u, u_{u\upsilon} = u_u \cdot f_\upsilon, i_u = \frac{3}{2}(i_{1u} \cdot f_u + i_{1\upsilon} \cdot f_\upsilon); \qquad (3)$$

1.2. Equations of the HB

$$u_u = u_{вu}, u_\upsilon = u_{в\upsilon}, f_{вu} = \frac{2 \cdot \sqrt{3}}{\pi} \cos \theta_i, \; f_{в\upsilon} = \frac{2 \cdot \sqrt{3}}{\pi} \sin \theta_i,$$
$$u_в = \frac{3}{2}(u_{вu} \cdot f_{вu} + u_{в\upsilon} \cdot f_{в\upsilon}), \; i_u = i_в \cdot f_{вu}, i_\upsilon = i_в \cdot f_{в\upsilon}; \qquad (4)$$

1.3. Equations of the PZPT

$$\frac{di_в}{dt} = \frac{u_в - u_u}{L_{рф}}, \; \frac{du_u}{dt} = \frac{i_c}{C_{бф}}, \; i_c = i_в - i_u; \qquad (5)$$

1.4. Equations of the AD

$$u_{1u} = \frac{d\psi_{1u}}{dt} - \omega_k \psi_{1\upsilon} + R_1 i_{1u}, \; u_{1\upsilon} = \frac{d\psi_{1\upsilon}}{dt} - \omega_k \psi_{1u} + R_1 i_{1\upsilon}, \; J\frac{d\omega_э}{dt} = M - M_c,$$
$$0 = \frac{d\psi_{2u}}{dt} - \omega_k \psi_{2\upsilon} - \omega_э \psi_{2\upsilon} + R_2 i_{2u}, \; 0 = \frac{d\psi_{2\upsilon}}{dt} + \omega_k \psi_{2u} - \omega_э \psi_{2u} + R_2 i_{2\upsilon}, \qquad (6)$$
$$M = \frac{3}{2} p_n \psi_{0u} i_{1\upsilon} - \frac{3}{2} p_n \psi_{0\upsilon} i_{1u};$$

where $u^*_u$, $u^*_\upsilon$-giving effects; $f_u$, $f_\upsilon$- modulation index; $U_0$ is the amplitude of the reference voltage; $u_u$ - the voltage of the power supply of the inverter; $i_u$ - the average current of the inverter; $u_{uu}$, $u_{u\upsilon}$- the average output voltage of the inverter; $i_{1u}$, $i_{1\upsilon}$- average inverter output currents; $u_u$,$u_\upsilon$, $i_u$,$i_\upsilon$ are the transformed voltage harmonics at the power input of the NV and the network currents; $f_{вu}$, $f_{в\upsilon}$ - transformed harmonics of switching functions HB; $\theta_i$ is the angle of rotation of the resulting network current vector, with respect to the phase axis A of the mains voltage; $u_в$, $i_в$- voltage and current at the output HB; $L_{рф}$ is the inductance of the smoothing reactor LC filter; $C_{бф}$ - capacitance battery capacitance capacitor;$i_c$ - current of the capacitor; $u_{1u}$, $u_{1\upsilon}$, $i_{1u}$, $i_{1\upsilon}$, $\psi_{1u}$, $\psi_{1\upsilon}$are respectively the transformed voltages and the total fluxes of the stator windings, $i_{2u}$, $i_{2\upsilon}$, $\psi_{2u}$, $\psi_{2\upsilon}$ are the transformed currents and the total flux linkages of the rotor winding; $\psi_{0u}$, $\psi_{0\upsilon}$ - main flux linkage; $p_n$is the number of pairs of poles.

The criteria for the similarity of a frequency electric drive (according to the rule of integral analogs) are obtained by reducing the equations to the dimensionless form:

*1.4.1. for the system of equations AIN:*

$$\pi_1 = \frac{0.5 \cdot \overset{*}{u_u}}{U_0 \cdot f_u}; \quad \pi_2 = \frac{0.5 \cdot \overset{*}{u_\upsilon}}{U_0 \cdot f_\upsilon};$$
$$\pi_4 = \frac{u_и \cdot f_\upsilon}{u_{и\upsilon}}; \quad \pi_5 = \frac{1.5 \cdot i_{1u} \cdot f_u}{i_и}; \quad \pi_6 = \frac{1.5 \cdot i_{1\upsilon} \cdot f_\upsilon}{i_и}; \qquad (7\text{-}11)$$

*1.4.2. for the system of equations HB:*

$$\pi_7 = \frac{u_{\text{в}u}}{u_u};\ \pi_8 = \frac{u_{\text{в}\upsilon}}{u_\upsilon};$$

$$\pi_9 = \frac{1.103 \cdot \cos\theta}{f_{\text{в}u}};\ \pi_{10} = \frac{1.103 \cdot \sin\theta}{f_{\text{в}\upsilon}};\ \pi_{11} = \frac{1.5 \cdot u_{\text{в}u} \cdot f_{\text{в}u}}{u_{\text{в}}};$$

$$\pi_{12} = \frac{1.5 \cdot u_{\text{в}\upsilon} \cdot f_{\text{в}\upsilon}}{u_{\text{в}}};\ \pi_{13} = \frac{i_{\text{в}} \cdot f_{\text{в}u}}{i_u};\ \pi_{14} = \frac{i_{\text{в}} \cdot f_{\text{в}\upsilon}}{i\upsilon}; \qquad (12\text{-}19)$$

*1.4.3. for the system of equations PZPT:*

$$\pi_{15} = \frac{u_{\text{в}} \cdot t}{L_{\text{рф}} \cdot i_{\text{в}}};\ \pi_{16} = \frac{u_{\text{и}} \cdot t}{L_{\text{рф}} \cdot i_{\text{в}}};$$

$$\pi_{17} = \frac{i_c \cdot t}{C_{\text{бф}} \cdot u_{\text{и}}};\ \pi_{19} = \frac{i_{\text{и}}}{i_c};\ \pi_{18} = \frac{i_{\text{в}}}{i_c}; \qquad (20\text{-}24)$$

*1.4.4. for the system of equations AD:*

$$\pi_{20} = \frac{\psi_{1u}}{R_1 i_{1u} t};\ \pi_{21} = \frac{u_{1u}}{R_1 i_{1u}};$$

$$\pi_{22} = \frac{\psi_{1\upsilon}}{R_1 i_{1\upsilon} t};\ \pi_{23} = \frac{u_{1\upsilon}}{R_1 i_{1\upsilon}};\quad \pi_{24} = \frac{\psi_{2u}}{R_2 i_{2u} t};\ \pi_{25} = \frac{\omega_э \psi_{2\upsilon}}{R_2 i_{2u}};$$

$$\pi_{26} = \frac{\psi_{2\upsilon}}{R_2 i_{2\upsilon} t};\quad \pi_{27} = \frac{\omega_э \psi_{2u}}{R_2 i_{2\upsilon}};\ \pi_{28} = \frac{p_n \psi_{0u} i_{1\upsilon}}{M};\ \pi_{29} = \frac{p_n \psi_{0\upsilon} i_{1u}}{M};$$

$$\pi_{30} = \frac{J\omega_э}{Mt};\ \pi_{31} = \frac{M_c}{M}. \qquad (25\text{-}36)$$

Similarity of processes in the original and similarity means that they, having qualitatively the same character and differing only in scale, should be described by the same mathematical equations. Obviously, the similarity object will also describe the processes in the frequency drive - in the event that the ratios of the scaling coefficients for the associated terms of the homogeneous equation are equal, an example for (20-24):

$$I_{\pi 1} = \frac{u_{\text{b}} \cdot t \cdot L'_{\text{pf}} \cdot i'_{\text{b}}}{u'_{\text{b}} \cdot t' \cdot L_{\text{pf}} \cdot i_{\text{b}}} = 1;\ I_{\pi 2} = \frac{u \cdot t \cdot L'_{\text{pf}} \cdot i'_{\text{b}}}{u'_{\text{и}} \cdot t' \cdot L_{\text{pf}} \cdot i_{\text{b}}} = 1;$$

$$I_{\pi 3} = \frac{i_{\text{c}} \cdot t \cdot C'_{\text{bf}} \cdot u'_{\text{и}}}{i'_{\text{c}} \cdot t' \cdot C_{\text{bf}} \cdot u_{\text{и}}} = 1;\ I_{\pi 4} = \frac{i_{\text{b}} \cdot i'_{\text{c}}}{i'_{\text{b}} \cdot i_{\text{c}}} = 1;\ I_{\pi 5} = \frac{i_{\text{и}} \cdot i'_{\text{c}}}{i'_{\text{и}} \cdot i_{\text{c}}} = 1. \qquad (37\text{-}41)$$

where parameters written with a stroke refer to a similarity model.

A feature of the similarity of such complex systems as a frequency electric drive consisting of subsystems HB, PZPT, AIN and AD, respectively, similar for the research electric drive and the industrial object, it is necessary that the similarity criteria for these subsystems be equal (7-41 ), which are composed of the parameters (3-6) common to these subsystems.

**Acknowledgments**
The article was prepared within development program of the Flagship Regional University on the basis of Belgorod State Technological University named after V.G. Shukhov.

**References**

[1] Vinogradov A, Zyabkina O, Mikhailova M 2013 Analysis of indicators of the quality of electrical energy in the power supply system of the Lebedinsky Mining and Processing Combine in the Belgorod Region *Bulletin of the Belgorod State Technological University named after V G Shukhov* vol 3 pp 175-180
[2] Averbukh M, Prasol D, Khvorostenko S 2017 Experimental study of non-sinusoidal regimes of the

shop system of power supply under dynamic vibrational formation of concrete mixtures*Electrical systems and complexes* vol 1 chapter 34 pp 24-31
[3] Averbukh M, Kuznetsov V, Korzhov D 2013 Problems of ensuring electromagnetic compatibility in electrical installations of industrial enterprises. *Bulletin of the Belgorod State Technological University named after V G Shukhov* vol 5 pp 203-208
[4] Josef Kuneš 2012 *Similarity and Modeling in Science andEngineering* (Czech Republic: Springer) p 49
[5] Braslavsky I, Ishmatov Z, Polyakov V 2004 *Energy-saving asynchronous electric drive*(Moscow: Academy) p 256
[6] Venikov V*Theory of similarity and modeling (for power engineering)*ed. 4 th, add. and pererab. (Moscow: LIBROKOM) p 479

# FLY ASH IMPACT FROM THERMAL POWER STATIONS ON THE ENVIRONMENT

**N I Kozhukhova[1], M S Lebedev[2], M I Vasilenko[3]and E N Goncharova[3]**

[1]Department of Materials Science and Materials Technology, Belgorod State TechnologicalUniversity named after V.G. Shukhov, Belgorod, 308012, Russia
[2]Department of Construction and Municipal Economy,Belgorod State TechnologicalUniversity named afterV.G. Shukhov, Belgorod, 308012, Russia
[3]Department of Industrial Ecology, Belgorod State TechnologicalUniversity named afterV.G. Shukhov, Belgorod, 308012, Russia

**Abstract**. Dumps of thermal power plants in Russia occupy significant areas. They are a source of pollution of air and water basins, contribute to increasing groundwater salinity and falling into the environment of radioactive elements. The work presents the results of experiments to identify the physical and chemical properties and toxicological effect of low-calcium ash from five thermal power plants on the environment.

One of the most significant sources the environment contamination are large-tonnage solid wastes from thermal power plants (TPPs). Therefore, in a number of regions these dumps have significantly complicated the ecological situation. Considering that a significant part of the country's electricity (about 70%) is produced by burning solid fuels, the increase in ash and slag wastes is expectedto be continue in the future. One of the less-demanded species of ash storage facilities is fly ash - fine-disperselow-calcium wastes obtained as a result of burning of coal fuel. They are characterized by inertness to water, therefore they can'tbe used as a hydraulic binding component in the construction industry. At present time, the examples of low-calcium fly ash are almost absent [1–3].The total volumeof utilization of these by-product does not exceed 7–10% [4]. Storage of accumulating fly ash in open areas may cause a negative impact on the ecosystems of nearby territories and the ecological situation as a whole [5].

Earlier, the authors studied different fly ash from Russian and foreign producer in terms of their availability as environmentally friendly raw materials [6, 7]. The biotesting methods used in the work demonstrated a positive or neutral effect on bioobjects, however, at the same time, significant differences in the effect depending on the producer of fly ashwere fixed, the reason of which have not been practically studied at present.

In this paper, the results of studies on the identification of physical and chemical characteristics and toxicological effect of fly ash on the environment were presented.

To determine the causes of toxicity of water extracts of fly ash, the determination of pH-value, total hardness, dry residue, which characterizes the presence of soluble salts in the filtrate, the composition of the residual ash after filtration. All listed indicators are environmental factors in watermedium, the effect of which on biological systems is the main factor.

The measurement of the pH-value was carried by ionomer pH-150M («Gomel Plant of Measuring Instruments»). The chemical composition of the dry residues from the aqueous extracts was determined withan energy-dispersive spectrometer based on a high-resolution scanning electron microscope TESCAN MIRA 3 LMU and a FTIR spectrometer VERTEX 70 (Bruker Optics) (absorption spectra of tablets of materials with potassium bromide in an average IR-region 370-4000 $cm^{-1}$).

Chemical composition of the studied fly ash before and after treatment in water medium as well as qualitative mineral composition of dry residue were determined with XRF-analysis using spectrometer ARL 9900 WorkStation («Thermo Fisher Scientific»).

The study was carried out for five low-calcium fly ash from Russian(1–3) and foreign (3, 4) thermal power plants. As a reference, the Portland Cement CEM 42.5N (Belgorodsky cement company, Russia) was used.

According to Russian Standard GOST 30108-94 andthe Norms of Radiation Safety (NRB-99/2009), a concentration of natural radionuclidein the used fly ashes islow and they can be use for all types of construction.

When measurement of pH-value of water extracts from fly ash, an increase in alkalinity was observed for all the samples vs. medium (pH> 8.23). This indicates that used fly ash contains soluble alkaline components varied in a wide range with a minimum content for fly ash 3 (pH 8.37), and maximal for Portland cement (pH 12.55). It is due to the presence of calcium hydroxide. The pH-value of the water extracts for fly ash 1, 2, 4, 5 is higher than for fly ash 1 (pH 9,00-9,33).

The degree of hardness of water extracts, i.e. the concentration of alkaline-earth metal ions in the composition of fly ash is capable of affecting the life activity of bioorganisms. The highest hardness of water extracts was observed for fly ash 4 and 5: 15.2–17.1 mg-equ/L. Other extracts from fly ash had a total hardness from 4.8 to 6.9 mg-equ/L.

Expect for the total content of salts, elemental composition is also important (Table 1).

**Table 1.**Elemental composition of dry residues from water extracts

| Fly ash | Content of chemical elements, weight % | | | | | | | | | | | |
|---|---|---|---|---|---|---|---|---|---|---|---|---|
| | O | Ca | Mg | Na | K | S | C | Cl | N | Si | F | Al |
| 1 | 19.8 | 5.0 | 2.6 | 16.1 | 3.2 | 4.7 | 16.1 | 31.8 | – | 0.6 | 0,1 | 0.1 |
| 2 | 35.3 | 7.1 | 6.4 | 5.0 | 5.3 | 6.1 | 10.3 | 19.8 | 1.5 | 2.6 | 0,4 | 0.1 |
| 3 | 36.8 | 7.1 | 5.8 | 6.1 | 3.7 | 5.0 | 8.7 | 21.3 | 3.4 | 1.6 | 0,6 | – |
| 4 | 45.7 | 15.7 | 0.7 | 4.0 | 1.7 | 14.1 | 13.0 | 4.3 | – | 0,7 | – | 0,1 |
| 5 | 40.5 | 10.8 | 0.5 | 7.0 | 1.3 | 9.7 | 22,3 | 7.2 | – | 0.2 | – | 0.1 |
| Reference | 11.7 | 16.4 | 0.1 | 5.4 | 10.7 | 1.5 | 9.3 | 44.8 | – | 0.1 | – | 0.1 |

The obtained results indicate the presence of elements such as $S^{6+}$, $C^{l-}$, $C^{4+}$ions, as well as $N^{6+}$, $F^{+7}$, $P^{+5}$ ions, which are typical for an acid residue.Also the presence of alkaline cations such as $Ca^{2+}$, $Mg^{2+}$, $Na^{+}$, $K^{+}$takes place. The high content of $Na^{+}$ and $Cl^{-}$ions negatively effect on life activity ofbioorganisms.

The significant differencein elemental composition of the dry residues is associated with different intensity ofions removal from the fly ash. In addition, the effect of atmospheric precipitation on dumps was simulated by long-term treatment in water medium of waste with distilled water.

According to the obtained results, significant decrease of $SO_3$and CaOin fly ash after washing was observed. (Table 2). Concentrationsofother oxidesarechangedslightlyordoesnotchanged.

The most significant decrease in the mass content of sulfur oxides occurs on fly ash 4 and 5. This is due to the presence of sulfates in the dry residue, which is confirmed by the data obtained (Fig. 1).

According to the results of IR-spectroscopy, for all dry residues, the content of sulfates, predominantly calcium, and various modifications thereof was observed, for example, gypsum ($CaSO_4 \cdot 2H_2O$), bassanite ($CaSO_4 \cdot 0.5H_2O$) - vibrations of $SO_4^{2-}$ groups with characteristic absorption bands at 601, 660, 668, 1096, 1118, 1140, 1150 $cm^{-1}$ and structural $OH^-$ groups with absorption bands at 1620, 1686, 3243, 3406, 3492, 3547, 3608 $cm^{-1}$) (Fig. 2). In Portland cement, carbonates were detected (peaks at 710, 874, 1405 $cm^{-1}$). The presence of chlorides with IR-spectroscopy is very difficult to detect.These salts are mostly transparent in the infrared range, they can be detected by sorbtive water, but the presence of other crystalline hydrates makes this task more complicated. However, they were detected with qualitative XRF-analysis (Fig. 2).

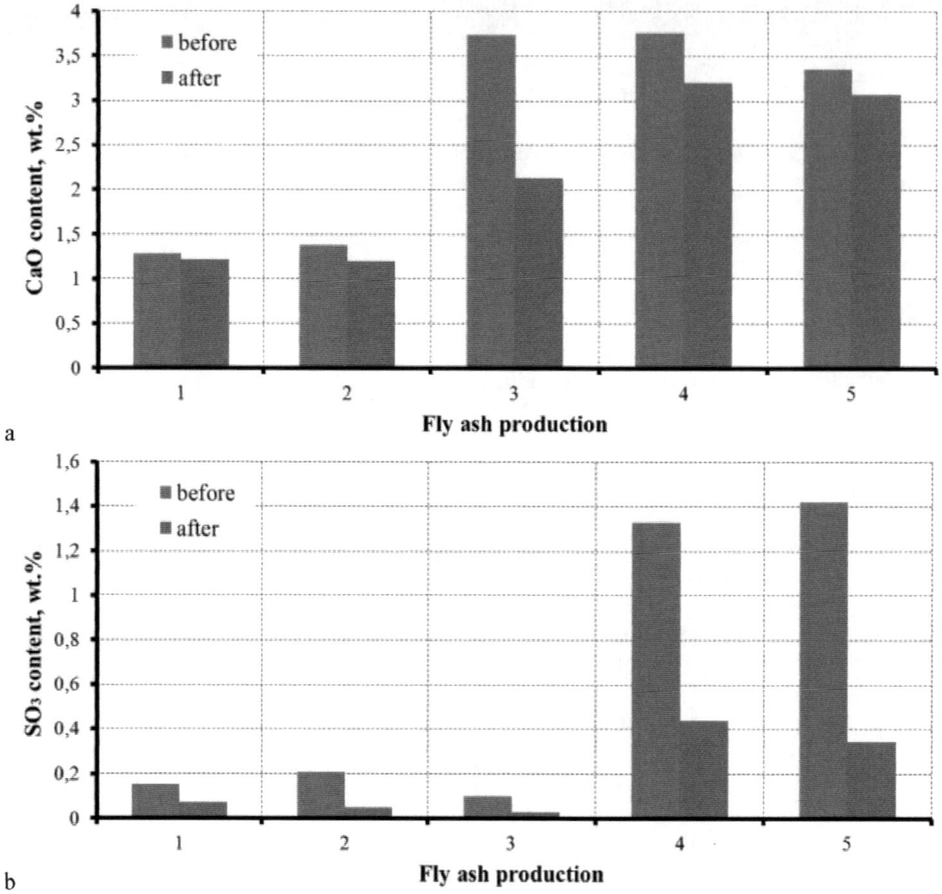


**Figure 1**. Content of CaOand $SO_3$ oxides in fly ash before and after long-term treatment in water medium

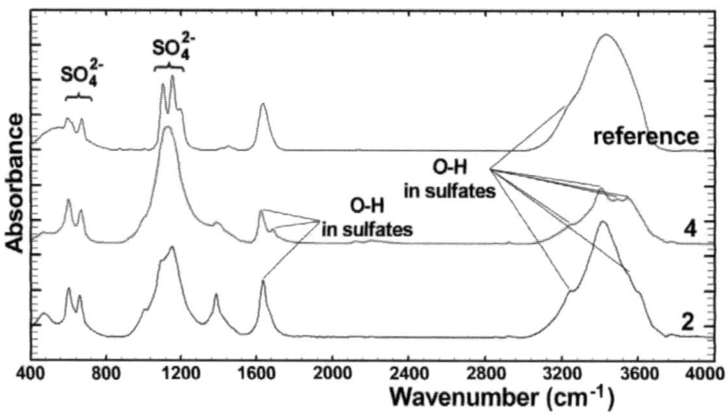


**Figure 2**. IR-spectra of dry residues of water extracts

In particular, halite is one of thecomponents in all water extracts. Potassium chlorides and magnesium were in samples of fly ash 1–3 were detected. In dry residuesof fly ash 4 and 5 (Fig.3), the predominance of sulfates, mainly gypsum, was observed. The appearance of the reflection forsassolite ($H_3BO_3$) is associated with a small amount of studied material for test and its extra contentwhen the sample preparation (Fig. 3).

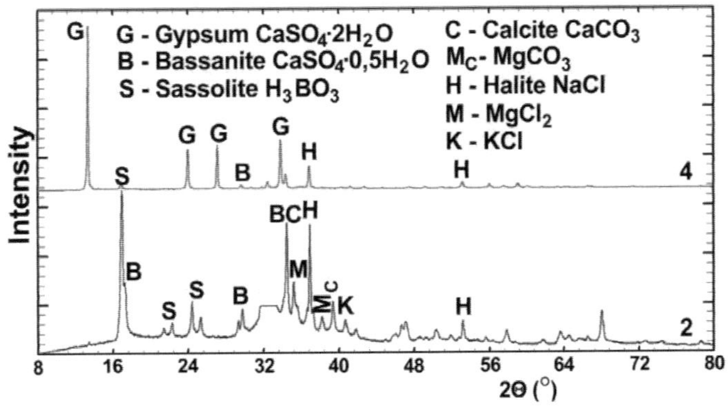


**Figure 3.** X-ray patterns of dry residues from water extracts

The dry residues from water extracts for fly ash 1–3 the crystals of sodium, magnesium and potassium chlorides, and less extent of carbonates and sulfatetake place (Fig. 3).Chemical composition of dry residues from fly ash 4 and 5 is characterized by the predominance of calcium sulfates and less of chlorides.

As shown in [5], under natural water exposure, the elements of fly ash those are negative for bioorganisms are removed from fly ash. According to the obtained results of biotest of water extracts, using Daphnia magna Straus crustaceans as the test objects for toxicity determination demonstratea dramatic increase inmortality of daphnia (more than 90%) with a reduction of their lifetime (not more than 24 hours). A high concentration of leached salts from water extracts of fly ash can initiate a high mortality of daphnia. Experiments carried outwith other test organisms demonstrated the same results.

Thus, as a result of the experiments was the fly ash from foreign producers (4, 5) more significantly affected the environment vs. fly ash from Russian producers (1–3).pH-valuefor water extract from fly ash 1–3 are close to neutral.The least amount of leached salts and average value of hardness are also typical for this fly ash.

**Acknowledgments**
This research work is financially supported by Ministry of Education and Science of Russian Federation in framework of State Assignment №11.9329.2017/8.9, using equipment of High Technology Center at BSTU named after V.G. Shoukhov.

**References**

[1] Belyakova Zh S, Velichko E G, Komar A G 2001 Ecological, material-lover and technological aspects of the application of ash-entrain of thermal power plants in concrete*Building materials***3** pp 46-48
[2] Goncharova E N, Porozhnyuk L A 2006 Investigation of the possibility of using waste as biocidal additives *Bulletin of BSTU named after V.G. Shukhov***1**pp 35-39
[3] Kozhukhova N I, ZhemovskyI V, StrokovaV V, Kalashnikova V A 2015 Influence of mechanical and chemoactivation processes on operational characteristics of geopolymerbinder*Res J ApplSci***10(10)**pp 620-623
[4] Nazirov R A, Kraft S L 2008 Experimental evaluation of leaching of natural radionuclides in the process of hydrosoldering of high-calcium ash. (*Izv. universitiesBuilding Novosibirsk: NSASU* )**1** pp 82-85
[5] Kraft S L 2008 The study of the problem of water resources pollution by natural radionuclides in the process of hydrosulphurisation of ash and slag wastes at thermal power stations (*Proceedings of the NSASU. Novosibirsk: NSASU*) **1** pp 148-152
[6] Kozhukhova N I, Zhernovsky I V, StrokovaVV 2015 Evaluation Of Geopolymer Binders Biopositivity Based On Low-Calcium Fly Ash*International Journal of Applied Engineering Research*vol**10(15)** pp 35527-35529
[7] Kozhukhova N I, Lebedev M S, Vasilenko M I, Goncharova E N 2006 Ecology-toxicology study of low-calcium solid wastes from power plant *Int. J. of Pharmacy & Technology***3** pp15349-15360

# HEAT INSULATING COMPOSITE MIXTURES WITH TECHNOGENIC MATERIALS

**L H Zagorodnyuk, A A Koryakina , K I Sevostyanova and A A Khaheleva**

Department of Building Materials Science, Products and Structures, Belgorod State Technological University of V.G. Shukhov, Belgorod, 308012, RU

**Abstract**. Rational use of techno genic materials with various physical and mechanical characteristics is one of the effective direction of resource-saving and the production of heat-insulating materials. The paper presents the results of experimental research and design development on the creation of a technological line for the production of heat insulation mixtures using techno genic porous aggregates.

At present, in Russia and abroad, the heat insulation of the exterior surfaces of buildings and structures is one of the most important tasks of resource-saving. Abroad, the greatest extension was obtained by heat insulation mixtures with porous fillers: fine-grained expanded clay, bloating glass granulate, perlite, vermiculite, spherical foam polystyrene, etc. [1].

The analysis shows that in Russia during in the operation of the communal facilities, up to 20% of all energy resources of the country are consumed, which is 2-3 times more than the energy consumed per unit area than in the European countries. In Europe during operation of multi-storey residential buildings, 350-550 kW / h of energy are consumed but cottages consumptionis 600-800 kW / h. At the same time, in the economically developed countries of Europe, residential buildings consume 90-120 kW / h of energy.

In this regard, the more active development of heat-shielding materials, along with the use of techno genic porous components (dust loss of burning aggregates of perlitic, vermiculite, ash components of TPP, etc.) in composite mixtures, as well as organic-mineral fiber-reinforced fillers, which have a reinforcing effect, is a very promising direction for science and practice.

At the same time, not only the urgent tasks of resource-saving are being solved, but also the rational use of techno genic materials, as well as the protection of the environment from pollution.

Heat insulation composite mixtures include a mixture of astringents, aggregates and various additives. Table. 1 presents the main components of dry construction mixtures.

**Table 1.**Components of dry building mixtures

| Astringents | Aggregatesandfillers | chemicaladditives |
|---|---|---|
| Portland cement, white cement, gypsum, anhydrite, lime, alumina cement, dispersiblepolymerpowders | Quartz sand, limestone, chalk, dolomite, perlite, kaolin, microsilica, ash loss, fibers, pigments, lightweight aggregates (expanded clay, bloating vermiculite and perlite, pumice, etc.) | Plasticizers, stabilizing and water-retaining additives dispersible polymer powders, retarders, accelerators, thickeners, pore-forming and antifoamingadditives. |

Porous fillers have a special place in heat-insulating composite mixtures (HIM).

In the studies carried out [2], the bloatingpearlitic sand produced by JSCOskolsnab (StaryOskol, Belgorod region) was used as a porous aggregate. The granulometric composition of the bloating perlitic sand is shown in Table. 2.

The chemical composition of perlite is represented by the following oxides, wt.%: SiO2 - 75.8; Al2O3 = 12.73; Fe2O3 - 1.47; CaO = 2.46.

X-ray phase analysis (XRA) showed that perlite sandis represented by such minerals ascristobalite, tridymite and feldspar.

**Table 2.**Granular composition of perlite sand

| Titleofresidue | Residuesonsieves, wt% | | | | | Passage through a sieve 0.14 wt% |
|---|---|---|---|---|---|---|
| | 2,5 | 1,25 | 0,63 | 0,315 | 0,14 | |
| particulate | 7,2 | 19,4 | 18,2 | 29,6 | 25,2 | 0,4 |
| Full | 7,2 | 26,6 | 44,8 | 74,4 | 99,6 | 100 |

When carrying out the experimental studies, the following HIM compositions were used table3.

**Table 3.**Blend composition of heat-insulating mixtures

| Compositionn umber | Ratio cement: bloating perlite sand by volume | V / T | Coneslump, cm | Average density, kg / m³ (28 days) | Average compressive strength, MPa (28 days) |
|---|---|---|---|---|---|
| 1 | 1:5 | 1,34 | 10,3 | 952 | 1,9 |
| 2 | 1:7 | 1,38 | 11 | 555 | 1,5 |
| 3 | 1:9 | 1,59 | 10,5 | 490 | 1,1 |
| 4 | 1:11 | 1,78 | 10,2 | 443 | 0,8 |
| 5 | 1:15 | 213 | 10,3 | 381 | 0,3 |

As a astrigent, portland cement CEM (PC 500 D0) was used. The water-hard ratio was determined according to the existing technical requirements for plaster solutions (mobility of the mixture OK = 10 ... 11 cm). The results of experimental studies (tests of cubes measuring 7.07 * 7.07 * 7.07 cm on compression when reaching the age of 28 days) are presented in the form of graphical dependence (Fig. 1). The ratio of cement-pearlite (C / P) varied in the range 1: 5 ... 1:15.

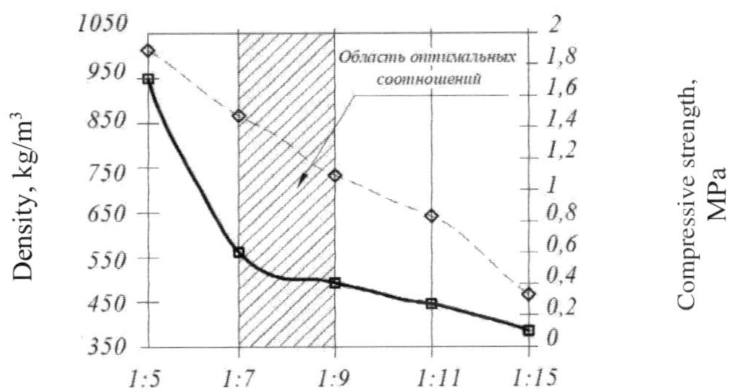


**Figure 1.** Dependence of the strength and density of samples on the ratio"Cement: pearlitic sand" of the composite mixture:
—— - density curve; - - - - – strength curve

The analysis shows that at a ratio of C / P= 1: 7, the following values are achieved: sample density $\rho$ = 550 kg / $m^3$, compressive strength $\sigma$ = 1.5 MPa (with the minimum permissible strength of plaster solutions $\sigma$ = 1,0 MPa).

The study of the influence of the granulometric composition of perlite sand on the physicomechanical characteristics of the samples showed that its selection is determined by the ultimate goal-reaching the maximum strength of the samples (at a higher density) or obtaining samples with increased heat-shielding properties.

Samples with grain sizes of 140 ... 630 μm (at 50% content, by weight) possessthe lowest thermal conductivity. Under increasing the content of large fractions, the density rise is observed and consequently, the mechanical strength with a higher heat conductivity.

In connection with the above said, we studied the possibility of using percolate dust in percolate production, pearlite dust trapped in cyclones and bag filters. Perlite dust is characterized by a high silica content ($SiO_2$ = 75.7%) and alumina ($Al_2O_3$ = 12.8%), as well as a greater absorption of Ca2 + ions from the saturated lime solution compared to the composite perlite.

Under the ratio C/P = 1: 9 with the addition of 20% perlite dust and the use of modified additives (Tylose, Tylovise SE, Hostapur OSB, etc.), the density of the plaster solution reached 490 kg / $m^3$, and the compressive strength $\sigma$ = 2.2 MPa. The coefficient of heat conductivity of the HIM corresponds to SNiP 11-3-79 $\lambda \leqslant 0.3$ W / (m · K) or the European standard for plaster solutions EN 998-1 for grades T1 - $\lambda$ = 0.1 W / (m · K) and T2 - $\lambda$ = 0.2 W / (m · K).

It is useful to use vermiculite and vermiculite dust caught in the system to ensure the heat-insulating properties(with the necessary strength characteristics) to the plaster mixtures, in addition to, artistic-decorative form Fig.2 system [3].

**Figure 2.** Microphotograph of vermiculite

To obtain heat-insulating compositional mixtures of different compositions using techno genic materials, the technological complex developed by us can be used (Fig. 3) [4, 5]. The complex makes it possible to implement various technological operations: storage, preparation of various components of the thermal insulation mix (astringent, porous aggregate, techno genic fibrous fillers, organic additives); their qualitative homogenization in a recirculation mixer; packing and storage of finished products. If necessary, mechanical activation of astringent and fibrous fillers to a microdispersed state in a centrifugal grinding mixer of selective grinding is provided.

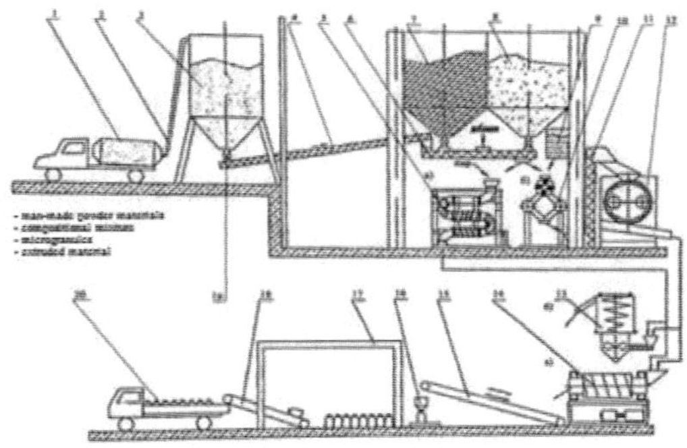

**Figure 3.** Diagram of heat-insulating composite mixtures:

1 - motor transport; 2 - pneumatic transport; 3 - bunker of astringents; 4 - auger conveyor; 5 - centrifugal grinding and mixing unit; 6 - auger feeder; 7 - bunker of porous aggregates; 8 - bunker of techno genic fiber-fillers; 9 - additive bunker; 10 - rotary-recirculating mixer; 15,18 - belt conveyor; 16 - weighing batcher; 17 - warehouse of heat-insulating composite mixtures of combined action; 19 - cell feeder; 20 - motor transport of finished products.

To conclude:
1. Investigations of physic and mechanical characteristics and the chemical compositions of the porous aggregates (perlite, vermiculite) and the dust collector caught during their production were carried out.
2. The kinetics of the process of gain in strength of perlite-containing compositional mixtures at different ratios of "cement-perlite" -1: 5 ... 1: 15 was found out.
3. Under the ratio C/ P = 1: 7, the density of the experimental samples reaches $\rho$ = 550 kg / m$^3$, and the compressive strength $\sigma$ = 1.0 MPa
4. Samples with size grains of perlite 140 ... 630 µm (at 50% of their content, by mass) possess the smallest heat conductivity. Under the ratio C / P = 1: 9 and adding 20% perlite dust and modified additives (Tylose, Tylovise SE, Hostapur OSB, etc.) to the composition the density of the plaster solution reached 490 kg / m$^3$, and the compressive strength $\sigma$ = 2.2 MPa.
5. To expand the technological and architectural-building capabilities of HIM it is rational to use vermiculite and its derivatives
6. A technological complex for the production of heat-insulating composite mixtures and special equipment for mechanical activation and homogenization of the mixture is developed

**Acknowledgments**
The article was prepared within development program of the Flagship Regional University on the basis of Belgorod State Technological University named after V.G. Shukhov.

**References**
[1] Zhukov A D, Stavritskaya L V, Sokov S D 2008 Heat insulation Materials, structures, technologies: info. allowance (Moscow: Stroyiform) p 440
[2] Shirina N V, Zagorodniuk L H 2010 Effective building mixtures for heat insulation works: Monograph *Belgorod: Publishing House of the Belarusian State Technical University* p184
[3] Zedgenizov V G, Nizhegorodov A I2010 A New Concept for the Development of the Domestic Vermiculite-Processing Industry *Modern technology, system analysis, modeling*vol 2 pp 100-103
[4] SevostyanovV S, UralV I, SevastyanovM V, NosovOA 2015 Technological complexes and equipment for processing and utilization of techno genic materials*Belgorod: BSTU Publishing House*p 321
[5] Glagolev S N, Sevostyanov V S, Gridchin A M 2014 Techniques and technologies for extrusion and drying of techno genic materials *Bulletin of Belgorod State Technological University named after V.G. Shukhov* vol 5 pp 118-123

IOP Conf. Series: Journal of Physics: Conf. Series **1066** (2018) 012012 doi:10.1088/1742-6596/1066/1/012012

# INCREASING OF THE ENERGY EFFICIENCY IN CIVILIAN BUILDINGS APPLYING FIXED TYPE OF FACADE SYSTEMS *

**M M Kosukhin , A M Kosukhin and K S Komarova**

Department of Construction and Municipal Economy, Belgorod State Technological University of V.G. Shukhov, Belgorod, 308012, RU

**Abstract**. The analysis of the technical condition of the fixed assets of the housing and communal sector of the country is presented, the main part of which are civil buildings. It is shown that the housing stock does not meet modern requirements of thermal engineering standards, energy saving and energy efficiency of buildings. To create the necessary microclimate of living quarters with a comfortable living environment, most civil buildings built at different times need modernization. To effectively carry out overhaul requires a comprehensive approach to the technology of repair and construction with the use of modern energy-saving technologies and materials. Effective implementation of thermal insulation of enclosing structures should be preceded by a detailed survey of the technical condition of facilities with the issuance of scientifically grounded recommendations based on modern scientific achievements in this field, for the further development of a technical task for the design of these works.

The main component of the housing and communal sector (HCS) of the country are civil buildings that represent the housing stock and social infrastructure facilities. At present, special requirements are required to create a comfortable living environment, meet the requirements of increasing energy efficiency and energy savings, increase operational reliability, and improve architectural expressiveness. Their implementation is a task of national importance and scale in the field of construction and housing and communal services of the country [1]. In addition, it is the largest sector of the Russian economy, which includes more than 26% of its total fixed assets. About 20 million objects of housing stock make up almost 3 billion square meters and consume more than 20% of the country's energy resources. The annual turnover in the housing and communal services sector exceeds 4,1 trillion rubles and is more than 5.7% of Russia's gross domestic product. Nevertheless, this is the most problematic branch of the economy. Depreciation of its fixed assets exceeds 60%, and in some respects it is higher. Among the problems hampering the development of the industry are: non-state system, administrative management, cost management, opacity of tariff formation, lack of public awareness, high percentage of moral and physical deterioration of communal infrastructure, and many others. But all these reasons can be attributed to subjective factors.

For the effective implementation of the tasks facing the industry, it is necessary to solve a set of measures that take into account its objective state. In the last decades of the reform of the housing and communal services, since the 1990s, numerous legal and regulatory acts have been developed in this area, including the "Strategy for the Development of the Housing and Communal Services of the Russian Federation for the Period to 2020" adopted January 26, 2016. The development strategy was supposed to contribute to improving the comfort of living conditions, upgrading and improving the energy efficiency of housing and communal services, the transition to the principles of the use of modern efficient materials and technologies used in the construction and modernization of public infrastructure and housing facilities, ensuring the accessibility of multi-apartment houses) for people with disabilities and other low-mobility groups.

But in connection with the fact that the approaches outlined in the strategy are mostly subjective in nature, the effectiveness of its implementation is far from expected [2]. Domestic experience clearly demonstrates the feasibility of using little-known and not currently used mechanisms and technologies. In work [3] the questions which are very actual for carrying out of reforms in a housing sphere and municipal (city) economy are opened.

The civil buildings currently under construction are more likely to meet energy efficiency requirements. As for the buildings built at different times, most of them do not meet the modern heat and technical requirements. Bringing them to the standard thermo-technical indicators requires major repairs and reconstruction, and, first of all, built according to the old thermal engineering standards. The desired effectiveness in this process can be achieved only by increasing the repair and construction industry itself, based on the results of a professional survey of the technical condition of civil buildings. To solve this task, an integrated approach to the technology of repair and construction works using modern energy-saving technologies is needed, and different from the traditional, generally accepted methods of organizing repair and construction.

Experience in carrying out major repairs showed that fragmented, unsystematic and local tasks did not solve the global problems of reducing the high energy consumption of buildings [4]. This led only to additional costs for such work, and the aging of the housing stock, its under-repair led to an increase in the amount of old and dilapidated housing and a decrease in the level of comfort.

When carrying out major repairs, in addition to ensuring the required parameters of the internal microclimate and operational reliability, a special role is assigned to improving the energy efficiency and energy saving of the renovated buildings. The main object of this goal is the enclosing structure. Since the publication of Federal Law No. 261-FZ "On Energy Saving and Energy Efficiency Improvement ...", almost all regions of Russia have begun to carry out major repairs of external fences of all types of buildings within the framework of implementing the requirements of this law. At the same time, the main indicators of the effectiveness of repair and construction work were the number of renovated houses and the volume of development of funds. The issues of quality and durability of the facade insulation stood in the background, and there was no scientific justification for such works at all. This was easily explained by the fact that the repair work should be preceded by a professional detailed examination of the technical condition of the facilities with the issuance of scientifically grounded recommendations based on modern scientific advances in the field of thermal insulation of enclosing structures for the further development of a technical design for the design of these works. But the payment for carrying out this kind of work has always remained a big problem.

Nevertheless, the work on efficient facade heat insulation requires a serious scientific approach with the aim of economical and technological efficiency of its choice and design. We must not forget that applied and fundamental scientific research is the direct source of innovative and breakthrough technologies, including in construction.

Enclosing structures that are exposed to the external and internal operating environment, is a complex thermodynamic system that obeys physicochemical laws, processes and phenomena. Therefore, it is possible to effectively manage its thermophysical properties only on a scientific basis. This is especially important when it comes to the use of new building technologies, materials and heat insulation systems. As already noted above, in the previously constructed residential houses, external enclosing structures were subjected to various technological methods to meet their thermal and technical parameters. But they did not completely solve the problems posed by modern thermophysics [5]. In this regard, almost all the enclosing structures require major repairs for their insulation.

For a correct understanding of the need for insulation of the enclosing structures of residential buildings, it is necessary to have a picture of the processes occurring under the influence of various factors both inside the building and outside.

The durability of the enclosing structures, and especially their thermophysical properties, is significantly influenced by climatic conditions: atmospheric precipitation, diurnal and seasonal temperature variations, wind loads, solar radiation, bio-corrosive processes, and processes and phenomena within structures, caused by these conditions. The influence of these factors leads to a decrease in operational reliability, both the enclosing structures and the whole building as a whole [6]. In the course of time, the thermophysical indices deteriorate noticeably.

The outer enclosing structure is the phase-separation surface when cold external and warm internal operating environments interact. At the same time, both on the surfaces of this border and inside it, new processes and phenomena occur that have a significant impact on the quality of thermal insulation, the internal microclimate of the premises, as well as on the operating floor.

Atmospheric precipitation (rain, snow, fog) leads to the moistening of external fences. Getting to the surface of the fence, the water through the open pores through the capillaries penetrates into the thickness of the fence, and can be found there in all three of its aggregate states: liquid, solid and gaseous (vapor).

When the temperature is lowered, a phase transition of water into ice occurs with an increase in volume by a factor of nine. This, in turn, leads to the formation of internal stresses in the material of the fence, causing mechanical damage. As the temperature rises, the reverse phase transition of ice into water occurs. The internal stresses decay, leaving in the body of the structure from the outside the originated network of microcracks from the previous freezing. Then the process repeats, and each new cycle contributes to greater penetration of moisture, and, consequently, to greater destruction. The greater the difference between diurnal and seasonal temperature fluctuations, the stronger and more often these processes occur (the mechanism of frosty destruction). The constant action on the wetted wall of the alternating load leads to its intensive destruction. It should be remembered that the moisture inside the enclosing structure in different quantities is always present, not only from the effects of atmospheric precipitation. It is subdivided into: building moisture, ground, hygroscopic and operational. Construction moisture is caused by the consequences of "wet" construction processes, its amount depends on the heating and ventilation systems. Ground moisture penetrates into the enclosure by capillary suction with poor vapor and waterproofing. Hygroscopic (sorption, vapor) moisture enters the interior of the structure due to the hygroscopicity of its material. A special role in the moistening of external fences is given to the operating moisture, which is often neglected. Nevertheless, water vapor is contained in the exhaled air of a person, a large number of them are formed from domestic and industrial activities, cleaning of premises, operation of water supply and sanitation systems. Excessive pressure causes water vapor to move outside the building. Part of the moisture leaves the room with the help of natural and forced ventilation systems, and some fall into the fences and try to get out. In this case, the steam condenses on cold surfaces and turns into condensed moisture. The place of this phase transition is usually called the dew point. Negative effects of humidification of external walls can be manifested not only from the effect of negative temperatures. Under the influence of sunlight and light, a favorable environment for the emergence and development of mold, rot, fungi and other microorganisms is created on the moistened fencing system inside and on the surface. This reduces the hygienic indices of the indoor microclimate of the premises and leads to negative consequences for the residents. It follows that the outer fences of civil buildings must meet a number of requirements. They should be cheap in erection and operation, strong and durable, fulfill their functional purpose, providing energy efficiency of the building with the required internal microclimate and being architecturally expressive.

To meet these requirements, there are currently various multi-layer facade insulation systems on the construction market. In the new construction there are three types of location of the insulation of the building: internal, directly in the bearing wall, external. When it comes to the insulation of already erected and inhabited houses during their overhaul, there is a question of ways of warming, from which side it is more effective to have a heater. Insulation from the outside is most suitable for residential buildings, as it has a number of advantages: the use of the effect of heat accumulation by the bearing structures of the building; removal of the "dew point" beyond the support structure; The useful volume of the room does not decrease.

In the reconstruction of old buildings, the most easily realized and effective is the external thermal insulation system, which is no less successfully used in new construction [7]. All of them are divided into plaster and ventilated, have their own design features, advantages and disadvantages.

Plastering refers to the systems of insulation of facades of fastened type or - the system of "wet" facades and is divided into light and heavy. The arrangement of facades of this type provides for lowering the cost of heating the building from 40 to 50%, reducing the loss of thermal energy through the enclosing structures and reducing the payment for thermal energy, improving indoor comfort in the premises, reducing emergency situations. Such facades play an important role in their construction on the real estate of the municipal formation, which leads to an improvement in the quality and reliability of heat supply, a reduction in fuel consumption, the release of additional heat capacity, an increase in the life of the housing stock, a reduction in tariffs for thermal energy.

Hinged facades of fastened type can be effectively used in industry, in the construction and improvement of private houses, common house systems, including. apartment buildings, social institutions (schools, hospitals, kindergartens, etc.), administrative and public buildings and structures.

Light plastering facade insulation systems are a multi-layer heat-insulating and decorative design in which the insulation is fixed with the help of adhesive composition and mehanic devices on the outer surface of the fence. Then the warming layer is covered with a protective and decorative layer of plaster, not more than 15 mm thick.

In general, the warming system "light wet method" allows to achieve a high-quality insulation of the building at a relatively low cost with high artistic and decorative expressiveness of the facade. However, the use of even the most expensive and high-quality materials does not guarantee the high quality of the work performed. Unfortunately, in the market of facade insulation you can meet companies whose "professionalism" can negate all the potentially high characteristics of such systems.

Heavy plastering facade systems consist of consecutively applied layers of insulation and plaster structures, carrying the functions in which the reinforcing mesh fulfills. In this case, the thickness of the plaster layers after the layer of thermal insulation, in contrast to light plastering systems, can reach 50 mm. A distinctive feature of such systems is that in them the insulation is not glued to the external wall, but is fixed by means of special dowels-anchors.

These systems have much longer service lives than light stucco facades. The guaranteed service life of some manufacturers is declared up to 50 years (light systems are guaranteed to serve for about 20 years). This has led to a very large number of their popularity in the Scandinavian countries. In our country they are just beginning to gain their authority.

The thickness of the thermal insulation of the outer wall should not be less than 100 mm. Costs for the thermal insulation of external walls are made up of the cost of materials (thermal insulation, plaster mixes, mesh, dowels, decorative finishing) and the cost of work. As practice shows, the difference in total costs when using thermal insulation thickness of 50 mm and 100 mm will be about 10% and the effectiveness will differ by one and a half times. Increasing the thickness of thermal insulation above 150 mm is not economically profitable. The total cost will grow much faster than the effect of saving heat. So, the optimal thickness of thermal insulation for the insulation of external walls (when using thermal insulation with a coefficient of thermal conductivity $\lambda \leq 0.041$ W / m·S) is 100 ... 150 mm.

Unlike other systems of external thermal protection of a building, in a system of thermal insulation of a "wet" type, the outer material can be applied practically of any kind, i.e. to reconstruct old buildings with preservation of architectural appearance, and in new construction to preserve the architecture of the whole district. The main difference between the "wet" type heat insulation from another popular thermal insulation system with the help of "ventilated facades" is the absence of an air layer.

The main advantages of the external thermal insulation of the building "wet" type: 1) are used in the reconstruction and new construction of any type of buildings; 2) there is practically no restriction on use; 3) ensure compliance with the norms and rules of the Russian Federation in the construction and operation of buildings; 4) the calculation of the thermal insulation system is not laborious, does not require much time (but requires a sufficient amount of experience, since the use of materials with incompatible properties will lead to additional unscheduled repair work); 5) improve the sound insulation of buildings; 6) installation is simple and does not require much time; 7) service life according to the manufacturers' data of 20-25 years; 8) cheaper in comparison with the system of "ventilated facades"; 9) low cost of repairs; 10) almost any materials can be used for facing the building; 11) reduces the heating of the house wall in hot weather, making the conditions in the rooms more comfortable; 12) in comparison with the "ventilated facade" the facade of the "wet" type does not create a draft of air, thus the system will not allow rapid spread of ignition.

At the same time, there are restrictions on the use of external thermal insulation of the building "wet type", namely: 1) the installation of systems can be performed only at an air temperature of more than 5 ° C, in the absence of external weather influences; 2) the plaster layer is sometimes exfoliated and damaged (since it does not have protection from external influence), it is necessary to monitor the building constantly, because the integrity of the thermal insulation layer may be impaired; 3) operation is allowed at an average daily minimum temperature of the coldest five-day period of the year not lower than 55 °C; 4) in the reconstruction of old buildings, damaged walls are difficult to dry from accumulated moisture and even; 5) the height of the building is not more than 75 m (15 floors).

The thermal insulation of the "wet type" facades is most often used in low-rise buildings up to 15 storeys. It should also take into account the fact that this system of thermal insulation practically does not react to wind currents, because of the low load-carrying capacity on the walls.

In the old housing stock under reconstruction, the insulation of the facades of buildings with painting or coating the outside with plaster is an economical way of finishing the building with thermal insulation and has no alternative to other similar products. The payback period depends on the material and height of the building, the thickness of the heat-insulating layer, the facing material, the manufacturer of the insulation, but do not exceed 5-10 years. the economy of thermal energy reaches 40-60%.

To prevent even the minimum accumulation of moisture between the heater and the wall, and to prevent the insulation from peeling off the wall during the freezing of thawing, it is necessary to correctly choose the thickness of the insulation in order to fully take the dew point to the insulation. This will solve the problem of the fungus in the wall.

Removal of the dew point in the heater, adherence to technology and proper performance of the work will lead to a high-quality warming of the walls, and will help create a comfortable microclimate in the house during the long period of operation of the thermal insulation system.

Due to limitations in the use of a thermal insulation system for high-rise buildings (a building over 15 floors high), it is practically not used, but is suitable for finishing loggias and balconies.

The main point to pay attention to when choosing a "wet" type insulation is the variety of manufacturing companies (there are more than 40 of them in Russia) and design organizations. Also, it should be borne in mind that this event in the reconstruction of the old housing stock is most effective in conjunction with glazing of loggias, balconies, installation of modern windows, temperature controllers and individual heat points, etc. The set of measures can give a huge economic effect, with a payback period of 15-25 years.

Thus, the foregoing allows us to conclude that the real increase in the energy efficiency of civil buildings in the process of thermal insulation of their facades can be achieved only by using scientific approaches to solving these problems. This will make it economically and technologically advantageous to choose the type of thermal insulation depending on the climatic location of the facility, the materials and technology of its device. The main thing in solving this problem is that not thermal engineering processes and phenomena affect the operation of the outer enclosing structure, and an effectively designed design spontaneously controlled these processes.

**Acknowledgments**
The article was prepared within development program of the Flagship Regional University on the basis of Belgorod State Technological University named after V.G. Shukhov.

*The article was prepared within the framework of the Base University Development Program on the basis of the BSTU named after V.G. Shukhov.

**References**

[1] Kosukhin M M 2016 From the history of creation to the modern state and prospects for the development of housing and communal services in Russia *Bulletin of the Belgorod State Technological University named after V.G. Shukhov* vol 12 pp 48-54

[2] Kosukhin M M, Kosukhin A M, Shevtsova A V, Bogacheva M A 2016 On the issue of optimizing the means of capital repair and reconstruction of housing stock taking into account the comfort of the living environment (*Science-intensive technologies and innovations: Intern. scientific-practical. Conf. Bel-city*) vol 2 pp 87-89

[3] Govorenkova T M, Savin D A 1999 Housing and leasing cooperation The experience of the new economic policy and the possibility of its application in modern Russia (Housing Almanac) vol 4 pp 98-106

[4] Sozinov P The overhaul was left without professionals

[5] Kosukhin M, Kosukhin A M, Sharapov O N, Bogacheva M A 2016 Energy saving in conditions of sustainable functioning, modernization and development of housing stock *Bulletin of the Belgorod State Technological University V.G. Shukhov* vol 10 pp 51-61

[6] Kosukhin M M, Kosukhin A M, Sharapov O N 2016 Ways to improve the energy efficiency of housing in the Belgorod Region during capital repairs *Science-intensive technologies and innovations: Intern. scientific-practical. Conf. Belgorod* vol 9 pp 55-62

[7] Kosukhin M M, Semak A V, Kosukhin A M 2016 Estimation of energy-saving potential on the basis of energy audit *Bulletin of the Belgorod State Technological University named after V.G. Shukhov* vol 12 pp 89-94

# INCREASING OF HEAT EXCHANGE EFFICIENCY IN CLINKER REFRIGERATOR

**V M Konovalov , A AKulakov , A V Leetovchenkoand A AGoncharov**

Department of Building Materials Science, Products and Structures, Belgorod State Technological University of V.G. Shukhov, Belgorod, 308012, RU

**Abstract**. One of the promising ways that significantly expands the technological possibilities of cement clinker production is the application of the method of chemical heat regeneration (HRT) and the use of gas synthesis (a mixture of H2 and CO) as the main process fuel. The influence of the use of chemical fuel regeneration in heat recovery systems (clinker coolers) on the activity of clinkers with a low content of the aluminum-ferrite phase is considered in the article.

The cement industry in Russia is a very energy-intensive industry, consuming more than 12 million tons of conventional top-oil annually. Taking into account the consumption of energy resources for the extraction of raw materials and its processing, it ranks third after the fuel and energy and metallurgical industries. In the production of cement by wet method, the share of costs for process fuel in its cost price exceeds 40%, and the unproductive fuel consumption is about 75% [1].

When improving the technology of cement production, designing new or upgrading existing furnaces, special attention should be paid to the processes of heat recovery and efficient use of secondary energy resources [2]. In this regard, the clinker cooler should be considered not only as an aggregate, the purpose of which is to cool the clinker, but primarily as a recuperator, providing an increase in the air capacity (its exergic potential) going to fuel combustion. With the development of a dry method of cement production and a reduction in the fuel and air consumption for its combustion, a limitation is established on the amount of heat that can be recovered by the refrigerator. Due to the limitation of heat recovery, there is a limit to which the clinker can be cooled. Existing types of clinker coolers, for cooling the clinker below 100 ° C at the request of the technological regulations, are forced to use more air than necessary to completely burn the fuel in the furnace. To maintain the clinker temperature in the range from 100 ° C to 50 ° C, 3 to 4 kg of air / kg of clinker is necessary. When returning all the air to the furnace as a secondary one, the efficiency of the refrigerator will approach 100%. However, the combustion process requires only 2 kg of air and the maximum theoretical efficiency will not exceed 86%. The implementation of practical recommendations for the rational distribution of cooling air in the sublattice space of the grate refrigerator type "Volga" at a number of cement plants ensured an increase in the thermal efficiency of the grate cooler from 0,7 to 0,93 and, as a result, a significant reduction in fuel consumption [3]. Since an increase in the efficiency of the refrigerator even by 1% leads to a decrease in the specific heat consumption of the entire furnace by 17-21 kJ / kg cl., Effective measures for the modernization of refrigerators still justify themselves.

The application of the method of chemical heat regeneration (HRT) in recuperative systems and the use of gas synthesis (a mixture of H2 and CO) as the main process fuel is one of the most promising ways that significantly increases the technological possibilities of cement clinker production and requires the solution of a number of scientific and applied tasks. In particular, the creation of thermochemical reactors (TCR) and the organization of the combustion of modified fuels.

Previous studies [1,4] have shown the possibility of a water-steam conversion of natural gas in the clinker layer at a temperature above 700 ° C. The heat of reaction has a significant effect on the amount of heat absorbed in the thermochemical reactor [5]. As a result of heating, evaporation and overheating of water and methane mixtures, with a stoichiometric ratio of components (0,47 kg CH4 and 0,53 kg H2O), the methane vapor conversion reaction changes the enthalpy of the initial components in the preparation stage (the so-called physical cold storage). In the temperature range from T = 273 K to T = 1000 K, it is ΔHphysis = 3.2 MJ / kg of the mixture.

$$CH_4 + H_2O = 3H_2 + CO - 207 \text{ kJ}$$

The chemical cold resource, equal to the heat of reaction, is $\Delta H_{him.} \approx 6.1$ MJ / kg of the mixture. The total cold resource of conversion of gaseous fuel will be $\Delta H\Sigma = \Delta H_{phys} + \Delta H_{him.} = 9{,}3$ MJ / kg mixture. For comparison, the change in the enthalpy of liquid hydrogen heated to the autoignition temperature (T = 850K) is $\Delta H_{phys} = 12$ MJ / kg. As can be seen, the hydrocarbon plus water composition that undergoes a physico-chemical conversion to TCP approaches the liquid cooling option by the available cold resource, and exceeds it by the relative cooling resource $\Delta H / H_u$ ($H_u$-heat of combustion). Thus, the physical and chemical utilization of heat and subsequent combustion of converted synthetic fuel lead to an increase in the efficiency of energy devices and fuel economy [7]. In addition, the burning of synthesis gas significantly reduces the amount of emissions into the atmosphere of carbon dioxide. Therefore, the considered TCP process can be considered an example of a new energy-efficient technology that reduces harmful impact on ecology.

However, the use of the TCP method in the heat recovery system in the clinker refrigerator, in addition to the technical task, suggests the study of the effect of a highly reducing medium on the phase composition of the clinker and the quality of the resulting cement.

The object of the study was clinkers of CJSC Belgorod Tsement and an experimental clinker synthesized from chemical reagents Al2O3, Fe2O3, SiO2 • nH2O, CaCO3 of the brand CHDA. The skilled clinker was distinguished by an increased content of iron oxide and had KH = 0,93; p = 0,7 and n = 1,7. The chemical composition of clinkers is given in Table 1. An increased proportion of calcium aluminoferrites should respond more clearly to changes in cooling conditions. The peculiarity of the experiment is that the clinkers undergo sharp cooling, in which the processes of crystallization of the liquid phase have passed, which should greatly slow down the possible diffusion of ions. The annealing temperature did not exceed 1100 ° C. For comparison, the other part of the samples was annealed at at the same temperature in the air.

**Table 1.**Chemical composition of raw mix

| Components | Composition, % | | | | | Total, % |
|---|---|---|---|---|---|---|
| | SiO2 | Al2O3 | Fe2O3 | CaO | MgO | |
| Experienced clinker | 13,92 | 3,37 | 4,82 | 43,49 | 0,00 | 100,00 |
| Clinker of CJSC Belgorodsky | 19,34 | 4,09 | 3,46 | 63,21 | 0,97 | 91,07 |

Synthesis of gas was obtained in the layer of test clinkers on a model installation. Butane and steam were blown through the layer for 10 minutes in a ratio of 1: 4. The composition of the gases from the reactor was estimated from the indications of the VARIO plus industrial gas analyzer. In the composition of gases, CO and N2 were fixed. Hydrogen was not determined. The presence of carbon monoxide indicates the reaction of steam-water conversion of hydrocarbons.

The phase composition of the cooling products was determined on a DRON-3 diffractometer. Diffractograms are shown in Fig. 1 and 2.

The air-cooled pilot clinker is represented in [5] by the main silicate phases: $C_3S$ (3,047; 2,78; 1,769 Å) , $C_2S$ (2,89; 2,75; 2,61; 2,18 Å), A-F the AF phase is predominantly$C_4AF$ (7,24; 2,65; 1,94 Å) и $C_3A$ (2,71; 1,91 Å). The presence of reflections (4,89; 2,69 Å) indicates the possible presence of the $C_{12}A_7$ phase.

In the experimental "iron" clinkers cooled in the HRT reactor, a certain decrease in the characteristic diffraction maxima of $C_3S$ (2,78; 2,61; 1,769 Å) and $C_3A$ (2,71 Å), and at the same time an increase in the intensity of reflection 2,648 Å. Reducing the interplanar distance from 2,657 Å to 2,649 Å indicates an increase in the share of $Al_2O_3$ as part of A-F phase, which makes it possible to identify this phase as $C_6A_2F$. In the clinker composition, in both cases a small amount of phase is fixed $C_{12}A_7$. At the same time, under reducing conditions when gas synthesis is processed, the intensity of reflections characteristic of Mayenite and $C_3A$ decreases. A distinctive feature of clinkers cooled in a gas synthesis medium is the appearance of a phase of free CaO (2,41; 1,70 Å).

a)

b)

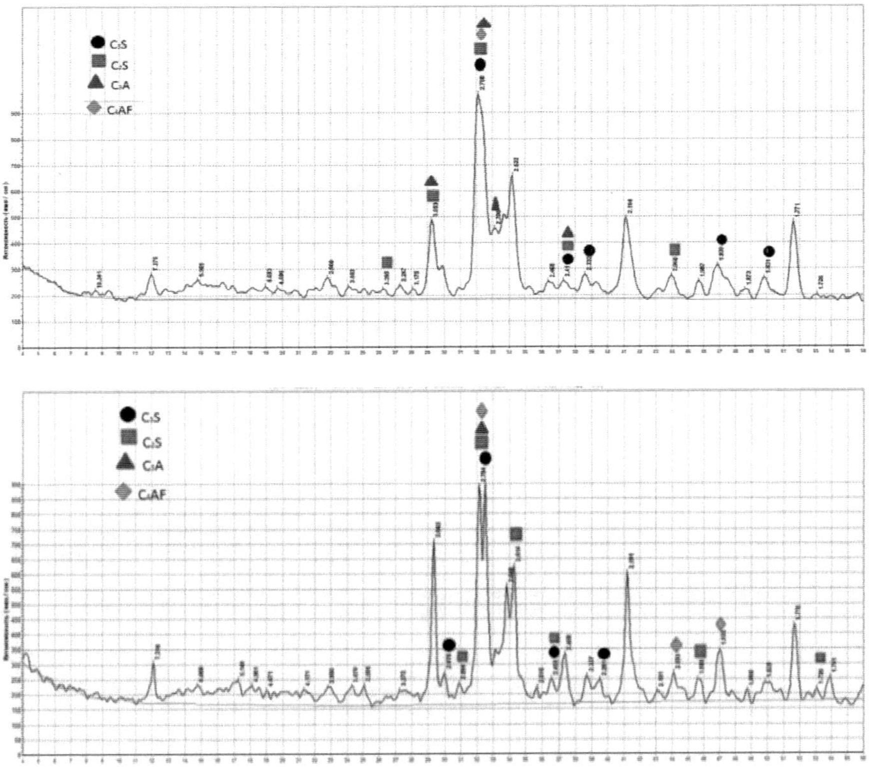

**Figure1**. Diffractograms of an experienced clinker:
a) cooled in air; b) in the HRT reactor

a)

b)

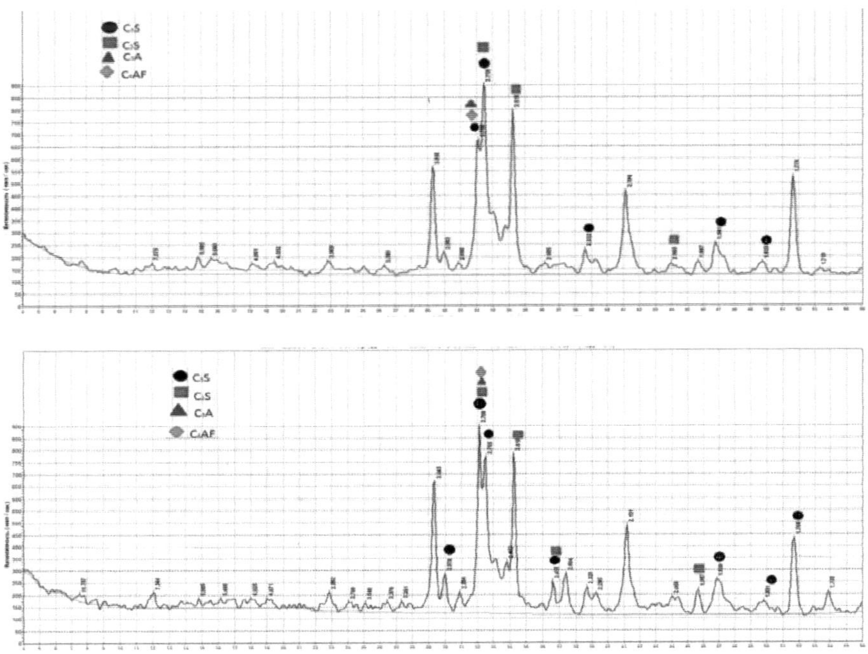

**Figure 2**. Diffractograms of clinkers of CJSC "Belgorodsky Cement" :
cooled in air; b) in a gas synthesis medium

Microscopic analysis of clinkers showed an increase in the bright intermediate substance, identified with the iron-containing phase, which indirectly confirms the data of the diffraction analysis.

Probably, the mechanism proceeding according to the reaction:

$$C_3A + C_4AF \rightarrow C_6A_2F + CaO.$$

Cooling of the clinker of CJSC"Belgorodsky Cement" in the gas synthesis environment (Figure 2) showed a slight decrease in the characteristic peak $C_3S$ (2,97; 2,76; 1,76 Å) и $C_3A$ (2,70 Å). At the same time, an increase in the intensity of the diffraction reflections of characteristic for β-$C_2S$ (2,78; 2,19 Å) and an increase in the amount of the alumino-ferrite phase represented by $C_4AF$ (7,24; 2,63 Å).

Cooling of this clinker in the gas synthesis medium was also shown by the appearance of unbound calcium oxide (2,41; 1,70 Å).

Microscopic analysis of apparent destruction of alite crystals did not show, but in the composition of the intermediate substance an increase in the fraction of the light phase was recorded, which can be identified as $C_4AF$. The obtained results suggest that the free CaO is released during phase decomposition $C_3A$.

Clinker Belgorodsky Cement CJSC, cooled under air conditions, is represented by the main phases: $C_3S$ (3,03; 2,76; 1,76 Å), $C_2S$ (2,89; 2,759; 2,61; 2,19 Å), $C_4AF$ (7,24; 2,63; 2,04 Å) и $C_3A$ (2,71; 1,91 Å).

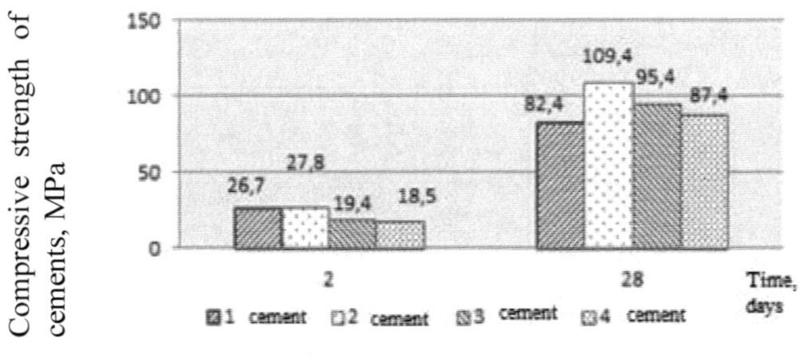


**Figure 3.** Compressive strength of cements:

1- cement of CJSC "Belgorod cement" and 2 - experimental cement cooled in air; 3 - Cement CJSC "Belgorod cement" and 4 - experimental cement, cooled in the synthesis gas

2-

Physical-mechanical tests (Figure 3) showed that the cement of CJSC "Belgorod cement" possesses the greatest activity in the early stages of hardening, and by 28 days the most durable was gained by the experimental cement cooled in air.

The results of the performed studies testify to the positive effect of the clinker cooling method in the reactor of chemical regeneration of fuel on the activity of clinkers with low content of the alumina-ferrite phase. For clinkers with a high content of iron oxides, cooling under reducing conditions leads to a significant change in the composition of calcium aluminoferrites, which is accompanied by the appearance of a new phase - $C_6A_2F$ release CaOand reduces the cement strength.

The use of chemical fuel regeneration in heat recovery systems makes it possible to use the heat of the cooled clinker more efficiently, and obtaining a significant amount of gas synthesis provides an increase in the exergic level of natural fuel, which increases the energy efficiency of production and reduces the cost of production.

## Acknowledgments

The article was prepared within development program of the Flagship Regional University on the basis of Belgorod State Technological University named after V.G. Shukhov.

## References

[1] Konovalov V M, Tkachev V V, Cherkasov AV 2016 Energy-efficient technology for cement clinker production: monograph *Belgorod: Publishing House of BSTU* p 96

[2] Klassen V K, 1993Cement clinker burning: monograph(Krasnoyarsk: Stroiizdat) p 320

[3] Konovalov S A,Pereskok M A, Petrova A N2016 Obrazov Improving the efficiency of heat and mass exchange processes in the production of cement *Bulletin of Belgorod State Technological University named after V.G. Shukhov* vol 4 pp 176-181

[4] Korabelnikov AV, Kuranov AL, Ryzhikov SS 2006 Physico-chemical kinetics in gas dynamics *Physico-chemical kinetics in gas dynamics* p12
[5] Mirkin LI 1976 X-ray diffraction analysis: reference manual *Moscow: Nauka* p 570
[6] Trubaev P A 2006 Exergy analysis of thermal processes in the building materials industry *Theoretical Foundations of Chemical Engineering* vol 40(2) pp 175-182
[7] Klassen V K, Trubaev P A, Novosyolov A G, Nuss M V. 2014 Sound management of the cement clinker burning with the use of information technology *International Journal of Applied Engineering Research* vol 9 pp 16855-16864

IOP Conf. Series: Journal of Physics: Conf. Series **1066** (2018) 012014 doi:10.1088/1742-6596/1066/1/012014

# INFLUENCE OF NON-SINUSOIDALITY OF VOLTAGE ON ELECTRICITY LOSS IN DISTRIBUTION NETWORKS

**M AAverbukh andM WAbdulwahhab**

Department of Power Engineering and Automation, Belgorod State Technological University of V.G. Shukhov, Belgorod, 308012, RU

**Abstract**. The problem of electromagnetic compatibility in high-voltage distribution networks is considered, connected with the consumption of reactive power at the fundamental frequency and distortion of the current form in the feeding networks. The estimated indices of the calculation of power losses for non-sinusoidal supply voltage in the high-voltage distribution network for the PNS pump with a capacity of 50 MW are given. Methods for reducing the voltage non-sinusoidality.

Powerful dynamic superchargers of a megawatt class are the basis of gas-oil pumping stations of main pipe-wires. The energy efficiency, reliability and environmental compatibility of the pipeline systems largely depend on the energy characteristics of the dynamic superchargers. At present, there is a steady trend towards the use of high-voltage frequency converters for regulated electric drives of dynamic superchargers. Being a non-linear load for a high-voltage distribution network, frequency converters distort the shape of the current in the supply line and thereby cause distortion of the supply voltage, which adversely affects other electrical receivers.

The problem of electromagnetic compatibility (EMC) in high-voltage distribution networks is associated with the consumption of reactive power at the main frequency and distortion of the current form in the supply networks. At the same time, generating higher harmonics of current cause additional losses of electricity, heating insulation and other negative effects.

Pumping of oil through main pipelines is carried out in accordance with the technological chain (Figure 1).

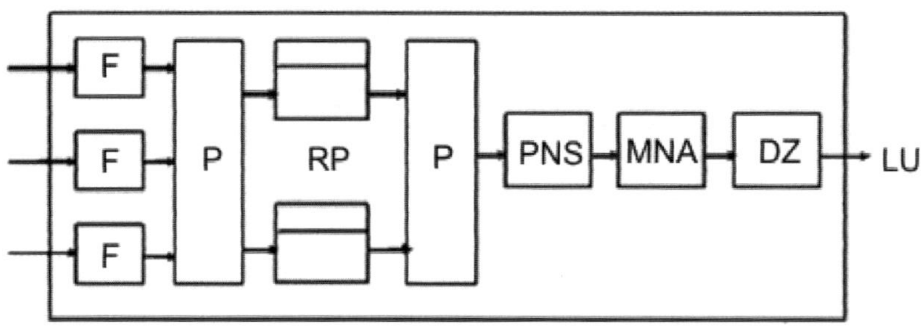


**Figure 1**. Technological process of oil transfer through the main pipe-line: F - receiving filters; P - the distributor; RP - tank farm; PNS - retaining pumping station; MNA - main pump unit; DZ - throttling gate; LU - linear section of the main pipeline

At the moment, mainly centrifugal pumps are used with drive high-voltage asynchronous motors with a squirrel-cage rotor. The installed power of one engine lies in the range from 3000 kW to 10,000 kW. Two methods are used to control the pressure and supply of oil: by controlling the damper at the pump outlet and by changing the rotor speed of the pump. The second method is the most energy efficient and can be used not only to regulate the rotor speed of the pump, but also to make the motor run smoothly, especially when the pump is running in a long-running nominal mode.

In Fig. 2 shows a fragment of the scheme of power supply of pump units with adjustable electric drives based on the frequency converter - asynchronous motor. It follows from Fig. 2, each pump can work when the motor is directly connected to a high-voltage network or through a thyristor high-voltage frequency converter that performs a soft start or regulates the rotational speed of the rotor of one of the pumps.

The higher harmonic components of voltages and currents in power supply systems cause additional losses of electrical energy in the supply lines, transformers [1, 2].

Calculation of power losses in power lines is made by the formula (1):

$$\Delta P_{\Sigma n} = 3\sum_{n=2}^{p} I_n^2 \cdot R_1 \cdot k_{rn} , \quad (1)$$

where n is the number of the harmonic; p is the number of harmonics taken into account; In is the current of the nth harmonic; $R_1$ is the line resistance at the fundamental frequency; $k_{rn}$ is a coefficient that takes into account the effect of the surface effect.

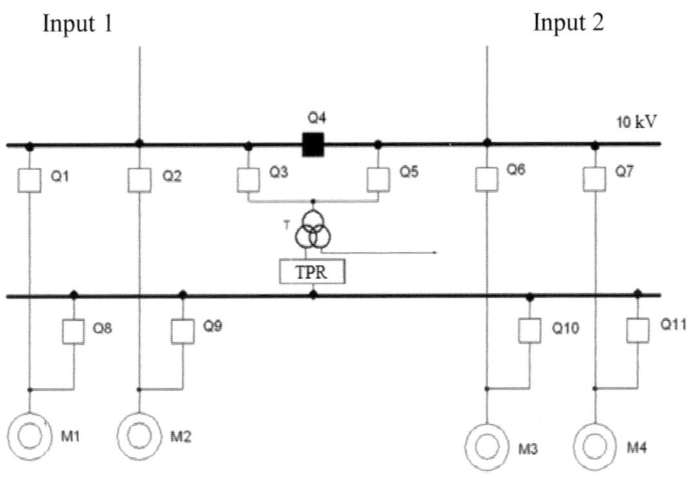


**Figure 2**. Scheme of power supply of pumping units of PNS

Calculation of power losses in the transformer is made by the formula (2):

$$\Delta P_{\Sigma n} = 3\sum_{n=2}^{p} I_n^2 \cdot R_{k1} \cdot k_{nm} , \quad (2)$$

where $R_{k1}$ is the short-circuit impedance of the transformer at the fundamental frequency; $k_{rn}$ is a coefficient that takes into account the increase in the short-circuit resistance for higher harmonics due to the effect of the surface effect and the proximity effect.

Preliminary calculations for a fragment of the power supply scheme show a significant influence of higher harmonics on the energy losses in distribution networks. The results of calculating the losses of power and electricity [4] in about eight hours of operation of one pump with a power of 5000 kW are presented in Table 1.

**Table 1.Estimated values for calculating power losses for non-sinusoidal supply voltage in the high-voltage distribution network ΔP, W, for the PNS pump, engine power 50 MW**

| | | | | | | | | |
|---|---|---|---|---|---|---|---|---|
| Losses in transmission lines | 1500 | 1100 | 1020 | 1350 | 1080 | 980 | 1320 | 1150 |

| | | | | | | | | |
|---|---|---|---|---|---|---|---|---|
| Losses in the transformer | 6700 | 6000 | 5900 | 6300 | 6150 | 5500 | 6200 | 6050 |
| Total losses | 8200 | 7100 | 6920 | 7650 | 7230 | 6480 | 7520 | 7200 |
| Total total power losses $\Delta W_{\Sigma tot}$,kW • h | | | | | | | | 466,4 |

The total additional power losses in the elements of the power supply system from non-sinusoidal power amounted to $\Delta P \Sigma n_{tot}$ = 58.3 kW, and the total additional power losses for seven hours amounted to $\Delta W_{\Sigma tot}$ = 466,4 kW ·h.

To reduce the voltage nonsinusoidality, it is possible by the following methods, which are divided into three groups:

- the scheme method: non-linear loads are allocated to a separate bus system; the location of this type of load on various nodes in the power supply system; grouping of converters according to the scheme of phase multiplication; connection of the load to a system with a power greater than that of a linear one:
- the use of narrow-band resonant filters, including their parallel load, as well as the inclusion of filtering, filtering, and active longitudinal and transverse inclusion filters;
- the use of equipment that have improved energy indicators: "non-saturated" transformers, multiphase transducers.

It should be noted that the methods of compensation are not universal: the choice of method depends on a number of factors [5]:

– level of generated harmonics, determined by the type of the source of higher harmonics (arc furnace, welding installations);
– the position of the source in the circuit and the power flows at the connection point and its effect on other electrical receivers installed in electrical proximity;
- availability of compensation means in the scheme
- frequency characteristics of electricity supply, as well as the probability of occurrence of resonance at the generated frequencies by a source of harmonics.

**Acknowledgments**
The article was prepared within development program of the Flagship Regional University on the basis of Belgorod State Technological University named after V.G. Shukhov.

**References**

[1] Zukhairi A., Vinogradov A 2014 *The status of the issue of use* (SMART GRID) vol 3 APRIORI Series: natural and technical sciences
[2] Averbukh M, Prasol D , Khvorostenko S 2017 Experimental estimation of parameters of regimes in high-voltage mine networks with powerful non-linear electric receiversvol 2*Vestnik of Irkutsk State Technical University* pp 75-84
[3] Dolinger S, Lyutarevich A, Goryunov V 2013 Estimation of additional power losses due to a decrease in the quality of electrical energy in the elements of the power supply system vol 2*Omsk Scientific Bulletin* pp 178-183
[4] Kartashev I, Tulsky V, Chamonov R (*Electricity quality management*vol 3*)*ed Sharov Yu. (Publishing house MPEI) p 347
[5] Averbukh M, Kuznetsov V, Korzhov D2013Problems of ensuring electromagnetic compatibility in electrical installations of industrial enterprises *Bulletin of Belgorod State Technological University named after V G Shukhov* vol 5 pp 203-207

IOP Conf. Series: Journal of Physics: Conf. Series **1066** (2018) 012015 doi:10.1088/1742-6596/1066/1/012015

# INVESTIGATION OF LANDFILL GAS OUTPUT FROM MUNICIPAL SOLID WASTE AT THE POLYGON

**P A Trubaev[1], O V Verevkin[2], B M Grishko[1], P N Tarasyuk[3], IISchekin[1], D Yu Suslov [4]and R S Ramazanov[4]**

[1]Department of Heat Engineering Technology, Belgorod State TechnologicalUniversity of V.G. Shukhov, Belgorod, 308012, RU
[2]Ltd TK Ekotrans, Belgorod, 308006, RU
[3] Ltd Energy Efficiency and Energy Saving, Belgorod, 308033, RU
[4]Department of Heat and Gas Supply and Ventilation, BelgorodState TechnologicalUniversity of V.G. Shukhov, Belgorod, 308012, RU

**Abstract**. The aim of the work is to study the output of landfill gas from the MSW landfill site, decommissioned 10 years ago, to determine the possibility of its energy use. The results of calculations of the output of landfill gas from the site according to two methods, the results of instrumental measurements of the bio gas composition, conducted during five months are presented. A series of measurements makes it possible to estimate the composition of the landfill gas being allocated in the absence of sludge: CH4 - 30 ... 32%; O2 - 0 ... 2%; CO2 - 40-60%; N2 - 5 ... 30%; H2S - 0 ... 30 mg / m3. Despite the lower methane content than in traditional bio gas, with regard to the energy potential, the landfill gas released at the experimental site allows it to be used provided that the combustion process is necessary.

The landfills, in which about 95% of all municipal solid waste (MSW) are located in Russia, are sources of pollution not only of soil and surface water, but also of the air basin [1]. In the body of the polygon under anaerobic conditions, as a result of the life activity of methanogenic bacteria, a landfill gas (bio gas) is formed, consisting of methane, carbon dioxide and nitrogen [2]. The greenhouse effect of methane is stronger than that of carbon dioxide, according to various estimates, from 25 times [3] to 84 times [4]. Therefore, at the waste treatment sites during operation and during their reclamation, measures should be envisaged for their degassing, which consists in the collection and utilization of landfill gas by incineration [5].

The aim of the work is to study the output of the landfill gas of the MSW polygon in Belgorod, operated by Ltd "TK Ekotrans", to determine the possibility of its energy use. Since projects on degassing of landfills are characterized by high cost, the work was provided with the maximum use of the production potential of Ltd "TK Ekotrans" in order to minimize the burden on consumers of services for removal and burial of MSW.

## 1. Calculation of the yield of MSW

To conduct degassing, part of the MSW landfill, located at st.Zarechenskaya, 85c, Streletskoe area, Belgorod region, Belgorod region (Figure 1) with the following characteristics:

Year of the beginning of functioning ............ 2001;
Year of the end offunctioning ........................ 2006;
The area of the site ....................................... 1,5 hectares;
Depth of ...................................................... 25 m;
The average volume of exported waste .......... 62 500 $m^3$ / year.

To calculate the composition of MSW was adopted as follows:

The content of the organic component in the waste ............. 55%

Organic waste content:........................................................
fatty substances............................................................. 2%
hydrocarbon-like substances.......................................... 83%
protein substances......................................................... 15%
Average Waste Humidity.................................................... 47%

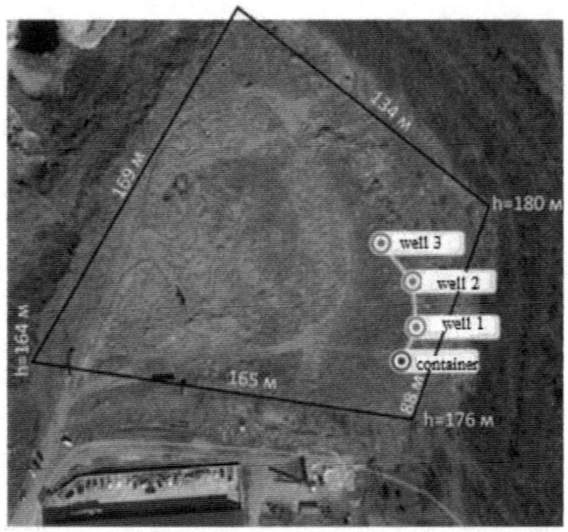


**Figure 1.** Landfill site for degassing with an experimental line of three wells

For the site, the theoretical yield of landfill gas was determined in accordance with [6] (procedure 1) and [7] (procedure 2), the results are shown in Fig. 2. The calculation of the bio gas yield for each method was carried out for a warm period and on an average for a year, the results obtained are consistent with other calculations given in [8, 9]. A period of active bio gas release was obtained for 17 years.

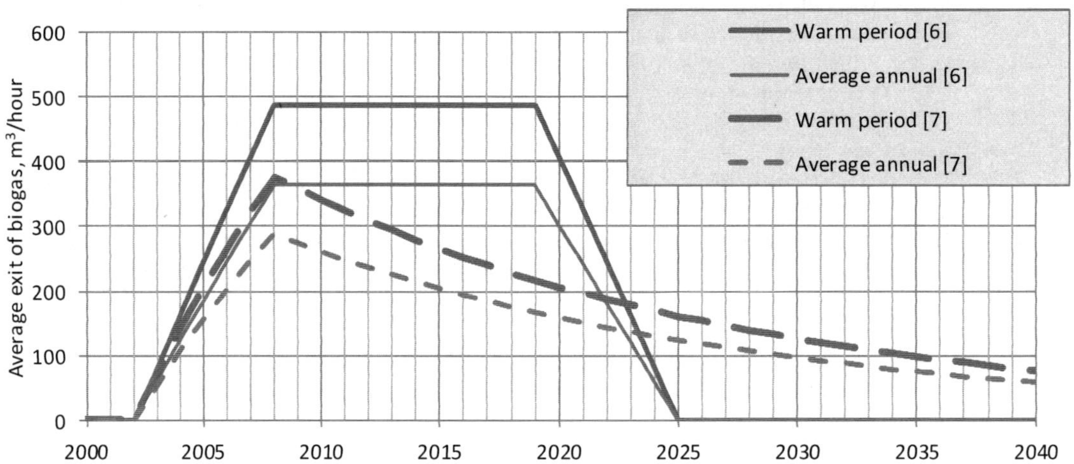


**Figure 2**. Forecast of bio gas output

## 2. Examination of the yield and composition of bio gas on the experimental line

For the study, three wells with a depth of 12 m (Fig. 3) were equipped with polyethylene perforated pipes with a diameter of 50 mm, produced by Ltd "TK Ekotrans" from polymer wastes (Figure 4).

IOP Conf. Series: Journal of Physics: Conf. Series **1066** (2018) 012015 doi:10.1088/1742-6596/1066/1/012015

**Figure 3.** Experimental well

**Figure 4.** Polyethylene perforated pipes of production Ltd "TK Ekotrans"

To measure the gas composition, a FARMEK FP34 gas analyzer with CH4 (0-100%), CO2 (0-2.5%), O2 (0-30%), H2S (0-100 mg / m3).

Measurements of the bio gas composition were initially made at the mouth of each well, while the results of the measurements were significantly affected by weather conditions - with increasing wind, the gas output from the wells increased significantly. In the period from June 20 to July 2, the site was sealed with clay, and three wells were combined with a gas gathering line. For gas removal at the end of the line, a non-oil plate-rotary vacuum pump Busch SecoSV 1005 D (flow rate 4.8 m3 / h, residual pressure 15 kPa) was installed in a specially equipped container).

The results of monitoring the composition of the landfill gas are shown in Fig. 5 and tab. 1. The composition of the gas is stable, after the discharge, the oxygen content does not exceed 5-10%, which shows that in the gas the oxygen impurities are not more than 25-50%. When the oxygen content is close to its theoretical content in the landfill gas (0-2%), the methane content is about 35% (Figure 6) The values obtained were less than the theoretical values (see Figure 1), which can be caused or transition of a polygon site to a stage of stable methano-genesis, or a difference in the MSW composition from the accepted.

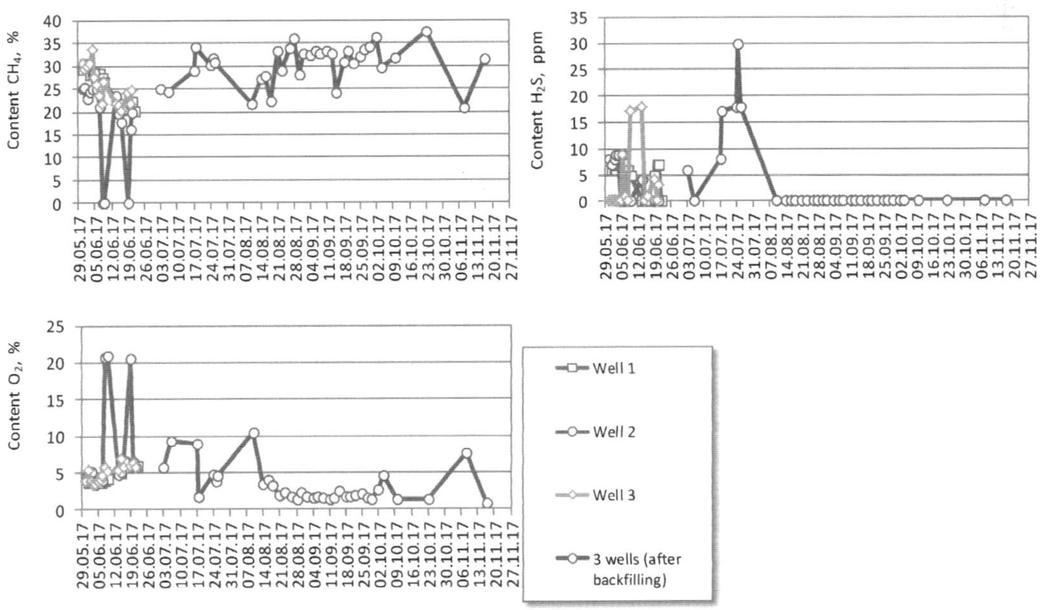


**Figure 5.** Monitoring of gas composition on the experimental line

**Table 1. The average composition of landfill gas after the merger of three wells into one and the back filling with soil (31.05-17.11.2017, 51 meters)**

| Value | $CH_4$, % | $O_2$, % | $H_2S$, mg / m$^3$ | $CO_2$, % | $N_2$, % |
|---|---|---|---|---|---|
| Minimum | 20,8 | 0,9 | 0 | – | – |
| Maximum | 36,2 | 10,5 | 30 | – | – |
| Average | **30,3** | **3,2** | **2,9** | – | – |
| Operation [7] | | | | | |
| clean biogas | 55 | 0 | 0–20 | 45 | 0 |
| biogas with air inflows | 35–45 | 1–6 | | 30–35 | 18–30 |
| Biogas composition of the Samara testing site (gas analyzer data) | 60,6 | 0,2 | 93 | – | – |

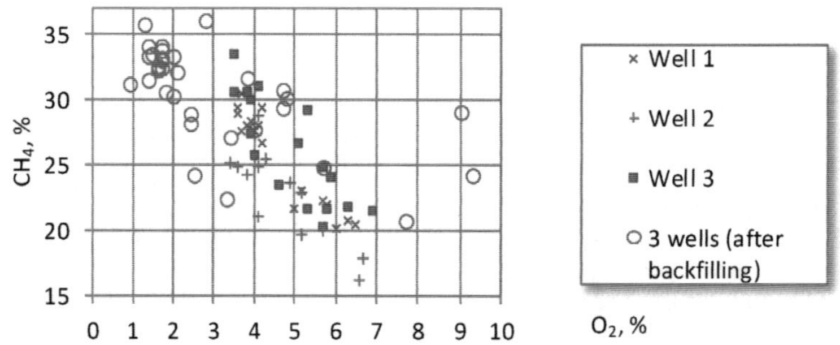


**Figure 6.** Effect of sludge on methane content

## 3.Determination of the energy potential of biogas

Utilization of biogas will be carried out by burning. Therefore, based on the data obtained, the calculation of combustion (tab. 2).

**Table 2.**Calculation of gas combustion ($\alpha$ = 1,2)

| Indicators | BiogaspolygonEcotrans | Naturalgas | Landfill gas of the polygon in Samara |
|---|---|---|---|
| Heat of combustion, kJ / m$^3$ | 10 740 | 36 322 | 21 695 |
| Airconsumption, m$^3$ / m$^3$ | 3,31 | 11,63 | 6,91 |
| Composition of combustion products, m$^3$ / m$^3$ | | | |
| $RO_2$ | 0,80 | 1,03 | 0,90 |
| $H_2O$ | 0,60 | 2,00 | 1,21 |
| $N_2$ | 2,80 | 9,19 | 5,56 |
| $O_2$ | 0,14 | 0,49 | 0,29 |
| Amount | 4,34 | 12,70 | 7,96 |
| Adiabaticcombustiontemperature, °C | 1487 | 1763 | 1666 |

The results show that for bio gas from the experimental site the heat of combustion and the air volume required for burning are two times lower than that of the gas of the Samara polygon and three times lower than that of natural gas. The burning temperature of these gases differs by no more than 20%. This shows that the landfill gas from the experimental section can be simply burned in the atmosphere (for converting

IOP Conf. Series: Journal of Physics: Conf. Series **1066** (2018) 012015 doi:10.1088/1742-6596/1066/1/012015

methane to a much less harmful CO2) [10], and is used in heat generation provided that the supply of the required amount of air is not more than the required volume for combustion.

According to the data of [11], when using bio gas as fuel, one has to face the following basic difficulties:

a) because of the high content of $CO_2$, the normal flame propagation velocity and its stability decrease, the burner control limits are narrowed;

b) because of the content of hydrogen sulfide, there is a problem of corrosion of burners, front plates of boilers, etc. (the service life of equipment is shortened).

Therefore, for low-calorie fuel, it is necessary to ensure good mixing with air, which is done with the help of swirlers or multi-channel burners [12].

## 4. Examination of the output of biogas from the production line

After confirming the possibility of burning bio gas on a part of the plot previously covered with clay, a bio gas collection line was constructed from 44 wells located 4 meters apart (Fig. 7). Wells were equipped with perforated pipes 50 mm in diameter, collectors for collecting bio gas are made of pipes with a diameter of 65 mm. After the lines were collected, a gas analysis was performed using the MRU Vario Plus gas analyzer equipped with sensors for determining the $O_2$, $CO_2$, $CH_4$ content (tab. 3).

**Table 3.Composition of biogas, % (17.11.2017 year)**

| Placeofsampling | $O_2$ | $CO_2$ | $CH_4$ |
|---|---|---|---|
| | 5,25 | 23,8 | 26,4 |
| Technological line (44 wells), in the absence of sludge | 0,99 | 38,3 | 32,8 |
| | 1,42 | 38,3 | 32,8 |
| Experimentalline (3 wells) | 0,85 | 57,8 | 32,8 |

**Figure 7.** A line for collecting biogas

The results of the measurements made it possible to determine the content of $CO_2$ in the landfill gas. Thus, the remainder in the landfill gas, ranging from 5 to 30% in different dimensions, and not determined by instrumental means, most likely consists of nitrogen $N_2$.

To conclude: A series of measurements of the bio gas composition of the experimental site, decommissioned 10 years ago, makes it possible to estimate the composition of the landfill gas being allocated in the absence of sludge: $CH_4$ - 30 ... 32%; $O_2$ - 0 ... 2%; $CO_2$ - 40-60%; $H_2S$ - 0 ... 30 mg / $m^3$. Supposedly, the remaining part in the gas, not determined by the instrumental method, $N_2$ - 5 ... 30%.

According to the energy potential, the landfill gas evolved in the experimental section is not inferior to natural gas or conventional gas with a large content of methane. But for sustainable combustion of gas it is necessary to ensure the supply of the appropriate amount of air, three times less than when burning natural gas, and intensive mixing of gas with air.

**Acknowledgments**

The article was prepared within development program of the Flagship Regional University on the basis of Belgorod State Technological University named after V.G. Shukhov.

**References**

[1] Shubov L Ya, Golubin A K, Devyatkin V V, Pogadaev S V 2000 The concept of solid household waste management (M .: GU NICPURO)p 72

[2] Kuschev L A, Suslov D Yu 2015 Theoretical description of the process of anaerobic fermentation in biogas plants *Bulletin of Belgorod State Technological University V G Shukhov* vol 6 pp 227-230

[3] Balachina T K 2012 Assessment of the impact of landfill gas from solid household waste polygons per person*Physiology Medicine Ecology of man* vol 2 pp 41-57

[4] Pachauri R K, Meyer L A 2015 Climate Change 2014: Synthesis *Report Contribution of Working Groups I, II and III to the Fifth Assessment Report of the Intergovernmental Panel on Climate Change* (IPCC, Geneva, Switzerland)p 151

[5] Kozhevnikov V P, Tokach Yu E, Ognev M N 2015 Modern solutions for the processing of solid domestic waste in BSTU V G Shukhova *Bulletin of Belgorod State Technological University named after V G Shukhov* vol 1pp 172-174

[6] Abramov N F, Sannikov E S, Rusakov N V 2004 Methodology for calculating the quantitative characteristics of pollutant emissions into the atmosphere from solid domestic and industrial waste landfills p 20

[7] Abramov N F, Vaisman Ya I, Maksimova S V 2003 Recommendations on the calculation of biogas formation and the choice of degassing systems on landfills for solid domestic waste disposal *Moscow: Federal State Unitary Enterprise Federal Center for Welfare and Waste Management* p 27

[8] Bespalov V I, Adamyan R G2013 Environmental safety of construction and urban economyDetermination of the amount of biogas generated at the landfill of solid consumer waste (*Naukovedenie*)vol 3 p 6

[9] Gorbovskaya A D, Khmish V H Kh 2016 Calculation of the formation of biogas in the utilization of waste at landfills *Actual problems of the humanities and natural sciences* pp 7-4, 20-24

[10] Kozhevnikov V P, Shirrime K J 2016 Ekologichesky aspect in industrial heat power engineering(Conf. Belgorod: Publishing House of BSTU) pp 95-103

[11] Khzmalyan D M, Kagan Ya A1976 Theory of combustion and combustion devices (Moscow: Energiya) p 487

# INVESTIGATION OF THE ALGORITHM OPTIMIZATION OF ELECTRIC POWER REGIMES ACCORDING TO VOLTAGE AND REACTIVE POWER WITH THE USE OF PENALTY FUNCTIONS METHOD

**A A Volovikov and A A Yudin**

Department of Power Engineering and Automation, Belgorod State Technological University of V.G. Shukhov, Belgorod, 308012, RU

**Abstract**. The article is devoted to the investigation of the algorithm for optimizing the steady-state regimes of electric power systems and networks for voltage and reactive power based on the application of the penalty function method. The results of the study are illustrated by the example of a 14-node test design scheme. Based on the analysis of the calculation results, conclusions are drawn about the effectiveness of this algorithm for optimizing the modes of closed-loop networks.

At present, when managing the modes of electric power systems and networks, the main attention is paid to ensuring reliable supply of electric energy of required quality to consumers in accordance with the Rules for the installation of electrical installations, the interstate standard GOST 32144-2013 "Electricity. Compatibility of technical means electromagnet. Norms of quality of electric energy in general-purpose power supply systems "and other normative documents. However, an equally important task is to minimize the costs [1] associated with ensuring the functioning of electric power systems and technological processes for the production, conversion, distribution and consumption of electricity [2], which acquires special significance after the adoption in 2009 of the Federal Law # 261- Fedaral Law "On Energy Saving and on Increasing Energy Efficiency and on Amending Certain Legislative Acts of the Russian Federation".

So, for example, in the calculation of network tariffs, the main attention is paid to the distribution of costs of electric grid companies by voltage classes, determining the losses of electric power and power, and distributing them among electric stations and consumers for the formation of prices. The issues under consideration are especially relevant for the energy sector, taking into account the new conditions for its market functioning and the use of modern information technologies [3]. Consequently, the study of algorithms for determining optimal steady-state regimes, that is, such admissible under the technical constraints regimes for which the optimization objective function takes the smallest value, is of great practical and theoretical importance.

The purpose of this paper is to investigate the algorithm for optimizing electric power regimes for voltage and reactive power based on the application of the penalty function method [3].

The optimization of the voltage and reactive power mode is used to determine the voltages at the nodes of the electrical network that are sources of reactive power. Most often, sources of reactive power are used to reduce the loss of active power and maintain the voltage within the required limits [4]:

$$U_{i\,\mathrm{min}} \leq U_i \leq U_{i\,\mathrm{max}}, \qquad (1)$$

where $U_i$ is the *i-th* monitored voltage value;

$U_{i\,min}$, $U_{i\,max}$ – minimum and maximum monitored voltage values.

Allowance for inequality constraints greatly complicates the optimization calculation. With the development of computer technology for solving optimization problems, methods of non-linear programming are widely used. Consider one of these methods, which is implemented in the software and computing complex RastrWin3, designed to solve problems in the calculation, analysis and optimization of the modes of electric power systems and networks [4].

It should also be noted that this software and computer complex allows you to perform optimization not only for voltage and reactive power, but also for the coefficients of transformation of regulated transformers and autotransformers. In this paper, this feature is not considered.

The calculation of the regime in the admissible region corresponds to the determination of the minimum of the objective function:

$$\Psi = \Delta P + \Sigma PF_i = \Sigma \Delta P_{ij} + K_j \Sigma \delta U_i^2, \tag{2}$$

where $\Delta P$ – total losses of active power;

$PF_i$ – penalty functions, which are introduced when the constraints (1) are violated by the voltages in the nodes;

$\Delta P_{ij}$ – active power losses in branch $i$–$j$;

$K_j$ – penalty coefficient, the value of which is chosen empirically;

$\delta U_i$ – violation of constraints (1) for stresses in nodes, which is defined as follows

$$\delta U_i = (U_i - U_{i\,\max})/U_{i\,\max}, \text{ if } U_i > U_{i\,\max};$$

$$\delta U_i = 0, \text{ if } U_{i\,\min} \le U_i \le U_{i\,\max};$$

$$\delta U_i = (U_i - U_{i\,\min})/U_{i\,\min}, \text{ if } U_i < U_{i\,\min}.$$

The losses of active power in the circuit branch can be calculated on the basis of the values of the modules $U_i$ , $U_j$ and the angles $\delta_i$ , $\delta_j$ of the voltage at the beginning and end of the branch $i$–$j$, respectively:

$$\Delta P_{ij} = (U_i^2 + U_j^2 + 2U_iU_j\cos(\delta_i - \delta_j))G_{ij},$$

where $G_{ij}$ – active conductivity of the branch $i$–$j$.

The minimum of the objective function (2) is determined by changing the voltage modules and the reactive power of the generating units – reactive power sources:

$$Q_{gi\,min} \le Q_{gi} \le Q_{gi\,\max}; \tag{3}$$

$$U_{gi\,min} \le U_{gi} \le U_{gi\,\max}. \tag{4}$$

The optimal values of the voltages $U_i$ , which are independent variables, are determined as a result of the iterative process.

At each step of the iteration process, the following quantities:

1. The admissible direction of the greatest decrease in the objective function (2):

$$\Delta S = \Delta U_g.$$

2. Direction of change of dependent variables ($\Delta Q$, $\Delta U$, $\Delta\delta$) when the independent variables change in the direction $\Delta S$.

3. From the conditions determined by the inequalities (1), (3), (4), the largest permissible step $\lambda_{\max}$ is calculated in the direction $\Delta S$.

4. The function $\Psi$ values are calculated in the following points: $\Psi(S)$, $\Psi(S+(\lambda_{\max}/2)\Delta S)$, $\Psi(S+\lambda_{\max}\Delta S)$. Is defined $\lambda_{opt}$ , which corresponds to the smallest value of the function $\Psi$ on a range $[0; \lambda_{\max}]$. If $\lambda_{opt} = 0$, then the step $\lambda_{\max}$ is divided in half and on a new segment is again determined $\lambda_{opt}$. The number of repetitions of the step division procedure does not exceed a given number, and if $\lambda_{opt}$ remains zero, the optimization stops.

5. If the restriction of the step was one of the constraints (that is, $\lambda_{opt} = \lambda_{\max}$), the set of independent variables is changed.

6. New values of variables $S' = S + \lambda_{opt}\Delta S$.

7. Power imbalances are determined, and, depending on their magnitude, a new steady state is calculated.

Optimization ends if the loss reduction between adjacent iterations with numbers $k$ and $(k + 1)$ and the value of the penalty component does not exceed the preset values $\varepsilon_1$ and $\varepsilon_2$:

$$|(\Delta P_{(k+1)} - \Delta P_k)/\Delta P_k| \le \varepsilon_1;$$

$$|((PF_{(k+1)} - PF_k)/PF_k| \le \varepsilon_2.$$

Consider the application of this algorithm on the modified 14-node IEEE test circuit [5], shown in Fig. 1, whose rated voltages are 220 and 110 kV. In this scheme, node No. 1 is taken as the balancing and basic one, nodes No. 2, 3, 6, 8 are generating, they are reactive power sources whose parameters can be optimized. The values of the mode parameters before optimization are shown in Fig. 1.

Ranges of reactive power variation of generating units are taken on the basis of available data on the possibilities of regulating reactive power in the nodes of the electric network. In accordance with GOST 32144-2013, positive and negative voltage deviations at the point of transmission of electrical energy should not exceed 10% of the rated or agreed voltage for 100% of the time interval in one week. Thus, the minimum and maximum permissible limits for voltage variation in the circuit under consideration are the values 198 and 242 kV (rated voltage 220 kV), 99 and 121 kV (rated voltage 110 kV).

Optimization of the electric power mode is performed in the RastrWin3 software-computing complex in accordance with the above algorithm. The results of calculating the optimum values of reactive power and voltage in the nodes of the electrical network are presented in Table 1.

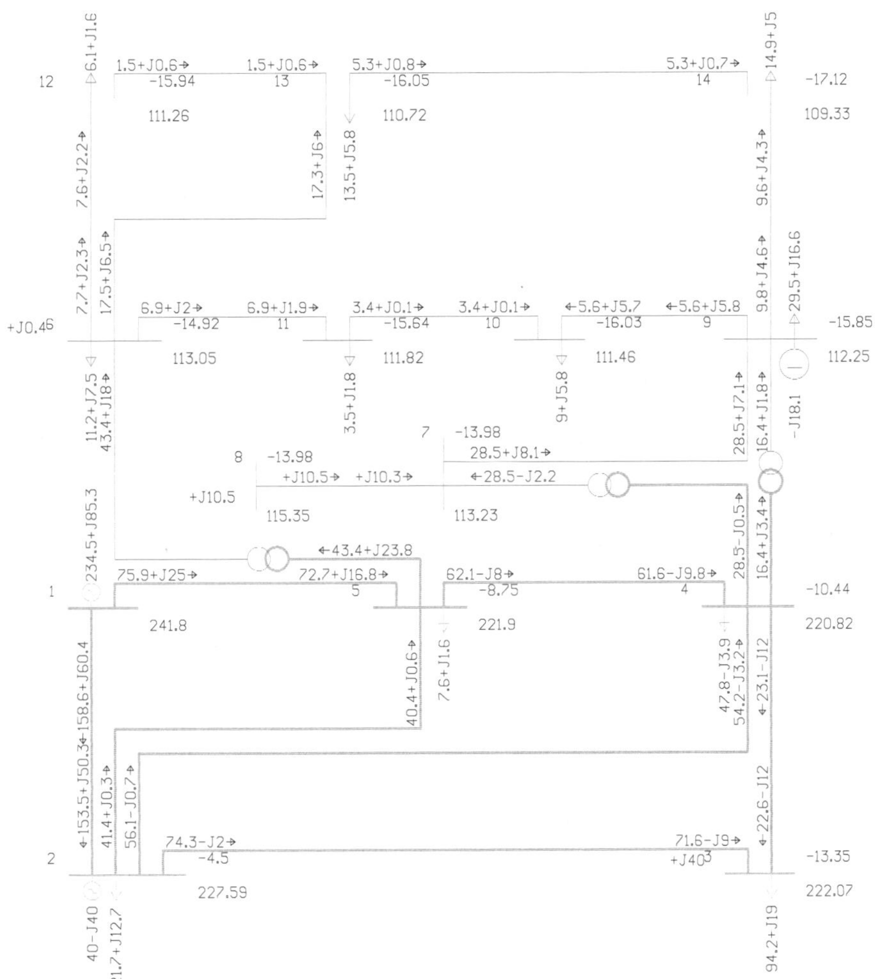


**Figure 1.** Test 14-node calculation scheme for algorithm analysis optimization in the software complex RastrWin3

The losses of active power in the branches of the scheme under consideration before optimization in the 220 kV network amounted to 14.86 MW, in the 110 kV network – to 0.62 MW. The losses of active power in the branches of the circuit under consideration after optimization in the 220 kV network amounted to 13.14 MW, in the 110 kV network – 0.58 MW. Thus, in general, according to the test scheme, the losses decreased by 11.37% compared to the original version. To calculate the optimum regime, it took two iterations.

**Table 1.The results of optimization of the electric network steady-state regime**

| Node number | $U_n$,kW | $U_{min}$, kW | $U_{max}$, kW | $U$, kW | $Q_{min}$, Mv | $Q_{max}$, Mv | $Q_g$, Mv |
|---|---|---|---|---|---|---|---|
| 1 | 220 | 198 | 242 | 241.80 | – | – | –9.4 |
| 2 | 220 | 198 | 242 | 238.07 | –40.0 | 50.0 | 44.8 |
| 3 | 220 | 198 | 242 | 230.49 | 0.0 | 40.0 | 30.0 |
| 4 | 220 | 198 | 242 | 230.88 | – | – | – |
| 5 | 220 | 198 | 242 | 231.47 | – | – | – |
| 6 | 110 | 99 | 121 | 121.00 | –6.0 | 24.0 | 13.1 |
| 7 | 110 | 99 | 121 | 119.19 | – | – | – |
| 8 | 110 | 99 | 121 | 121.00 | –6.0 | 24.0 | 9.4 |
| 9 | 110 | 99 | 121 | 118.81 | – | – | – |
| 10 | 110 | 99 | 121 | 118.32 | – | – | – |
| 11 | 110 | 99 | 121 | 119.24 | – | – | – |
| 12 | 110 | 99 | 121 | 119.22 | – | – | – |
| 13 | 110 | 99 | 121 | 118.62 | – | – | – |
| 14 | 110 | 99 | 121 | 116.61 | – | – | – |

Thus, based on the analysis of the results obtained, it can be concluded that the application of this algorithm provides an acceptable rate of convergence of the iterative calculations. In addition, this method has a high efficiency for optimizing the modes of closed-loop electrical networks.

**Acknowledgments**
The article was prepared within development program of the Flagship Regional University on the basis of Belgorod State Technological University named after V.G. Shukhov.

**References**

[1] Dubenko Yu and Dyshkant E 2016 Intelligent system of control and forecasting of energy losses *Bulletin of BSTU named after V G Shukhov* vol 3 pp 131-135

[2] Bakalin Yu, Mukhin N, Vinoglyadov V 2013 Restraining factors in the organization of energy conservation and issues of obtaining real energy efficiency Bulletin of BSTU V G Shukhov vol 2 pp 185-187

[3] Filippova T, Sidorkin Yu, Rusina A 2016 Optimization of regimes of power plants and power systems: a textbook ( Novosibirsk: Publishing house of the NSTU) p 356

[4] RastrWin3 - User documentation [Site]: The software complex "RastrWin" 2017 URL: http://www.rastrwin.ru/download/Files/

[5] Test schemes [Site]: Energy Department of the Institute of Energy and Power Engineering of the Komi Scientific Center of the Urals Branch of the Russian Academy of Sciences 2017 URL: http://energy.komisc.ru/dev/test_cases

IOP Conf. Series: Journal of Physics: Conf. Series **1066** (2018) 012017 doi:10.1088/1742-6596/1066/1/012017

# MATHEMATICAL MODEL OF ESTIMATION OF THE QUALITY OF THE APPLICATION OF THE CYLINDER-PISTON GROUP OF THERMAL DIESELS ON THE CRITERIA FOR THE STABILIZATION OF FUEL CONSUMPTION AT THE REAL TIME OF FACTORY TUBE TESTS

**V D Zonov and V P Kozhevnikov**

Department of Heat Engineering Technology, Belgorod State Technological University of V.G. Shukhov, Belgorod, 308012, RU

**Abstract.** In the article, a mathematical model of the evaluation of the quality of running-in of a cylinder-piston group (GPG) based on the criterion for stabilizing fuel consumption in real time for running tests of diesel locomotives on the diesel locomotive frame is considered. The effect on the magnitude of fuel consumption of suddenness is shown to be the inclusion of the disconnection of auxiliary equipment connected to the engine crankshaft via a V-belt transmission and a reducer with a compressor and water and air cooling fans. It was noted that the mathematical modeling of the on-off suddenness, carried out on the basis of the law of normal distribution, did not reveal a significant effect on the criterion of fuel consumption stabilization, with the help of which the quality of run-in was measured.

The influence of the critical state of the precision surfaces of the GPG on the fuel economy of diesel locomotives in operation is considered. The problem of critical wear of the precision surfaces of the cylinder head is considered. It affects the stable working process of the heat engine at practically all speeds of rotation and load, including idle and low load modes (non-nominal modes).

The solution of the problem of the stable working process of the heat engine is constrained by the absence of a theoretically grounded and experimentally confirmed criterion for stabilizing fuel consumption that estimates the quality of the run-in of the cylinder piston group in real time for each mode of shortened run-in tests of the internal combustion engine over the entire range of rotation and load frequencies.

The developed mathematical model presents the main performance indicators of the running-in of the GPG, which ensure a stable workflow in the process of the internal combustion engine. Requirements and conditions for qualitative technology of shortened run-up tests are formulated. A qualitative estimation of the work of the GPG in real time at all speeds and power of the diesel engine is given. The proposed mathematical model is maximally approximated to its use in engineering calculations by the engineering and technical staff of locomotive repair enterprises and locomotive depots.

Key words: fuel consumption stabilization criterion, shortened run-off tests of the GPG, mathematical model, precision surface, wear, diesel, workflow.

The resulted mathematical model allows to estimate quality of running-in of cylinder-piston group by criterion of stabilization of fuel consumption in real time of running tests of diesel locomotives on a diesel locomotive frame. The effect on the magnitude of fuel consumption of suddenness is shown to be the inclusion of the disconnection of auxiliary equipment connected to the engine crankshaft via a V-belt transmission and a reducer with a compressor and water and air cooling fans. It was noted that the mathematical modeling of the on-off suddenness, carried out on the basis of the law of normal distribution, did not reveal a significant effect on the criterion of stabilization of fuel consumption with the help of which the quality of running-in of the GPG was assessed.

When carrying out factory running tests of diesel generators on the diesel locomotive frame, the problem arises of the choice of the criterion for the quality of the running-in of the friction pair in the piston ring-cylinder liner at each run-in test mode in real time.

For the first time, the problem of a well-founded choice of the nomenclature of quality and reliability indicators was put in [1], where it was shown that the solution of this problem should be sought in the way of studying models of quality diagnostics and the influence of quality and reliability indicators on the quantitative characteristics of the process of functioning. In this work, as a general performance evaluation, the mathematical expectation of the income from exploitation is accepted. At the same time, income is understood as a useful return, which can be expressed in monetary or other form. The idea of the possibility of such an approach in solving problems related to ensuring reliability was expressed in [2]. Since the income from exploitation depends on the level of quality and reliability of the product, the selected indicators included in the expression for determining the expected income expectation were called the main ones. It is noted that they should be compared on the quality and reliability of different technologies and technical objects of the same purpose and different versions of the same product.

The analysis of various factors on which the qualitative nature of the work of the product depends, and those that determine the form of the income function and, consequently, the magnitude of its mathematical expectation are singled out.

As the main factors influencing the choice of the type of reliability indicators, it is determined:

Constructive solution factors (adaptability to repair);
• factors of the nature and mode of use of the product for its intended use;
• factors of failure consequences.

By factors of constructive execution of the product are divided into repaired and unreinforced.

According to the principles of limiting the duration of operation, the unmounted products are divided into the following:
• to failure;
• to failure or limit state;
• until the required functions are completed.

Repaired products according to the principles of limiting the duration of operation are divided into the following:
• until the first failure;
• until the first failure or limit state;
• until the first failure or completion of the required functions;
• to the limit state;
• to the limit state in the "standby" mode or to failure or perform the required functions in the "work" mode.

The division of products into repaired and non-repaired is due to the presence of an objective property embedded in the products during the design, regardless of whether the product can be repaired during operation or not.

According to the temporary mode of use, unremovable and repairable products are divided into operating

- in continuous mode;
- in discontinuous mode by regular cycles;
- in intermittent mode by cycles of random duration.

On the event of failure in the process of running-in the cylinder-piston group of an internal combustion engine, the most important is the fact of failure

when the specified functions are executed or not in the specified volume.

When assessing the functional consequences in the process of running-in the cylinder-piston group of an internal combustion engine, the failure events are distinguished by:

-fact of refusal;
- the performance or non-performance of the specified functions by the product in a given volume;
- forced downtime;
- refusal and forced downtime;
- the performance or non-performance of the specified functions in a given volume at an arbitrary time when the "work" begins.

The criterion for choosing a fuel consumption stabilization factor in real time for a running test of an internal combustion engine is to evaluate the functional consequences of failure, which can be estimated by the costs of eliminating the failure and, accordingly, the damage caused.

For all possible combinations of factors characterizing the real modes of running tests, formulas are used to calculate the economic efficiency during the operation of the engine and determine the reliability indicators corresponding to almost any diesel engine.

In accordance with the methodology of shortened run tests, the most significant factors characterizing the running conditions of the cylinder-piston group of the diesel engine on the diesel locomotive frame are determined. In addition, a cipher is compiled, according to which the indices of reliability and reliability are selected. The issue of selection of the quality indicators of the running-in of a cylinder-piston group is considered in accordance with the approved methodology of reduced running tests.

Further development of the method of factory shortened running tests, within the limits of the reduction in the running tests, is possible, in obtaining qualitative parameters of running-in of a pair of friction, a piston ring-cylinder liner with regard to efficiency in operation.

The choice of quality and reliability indicators of technology of reduced running tests of diesel engines is paid the closest attention, since possible failures, for example - jamming of the cylinder-piston group (CPG) failures of the department lead to a total failure of all diesel locomotive systems, thus reducing the operational efficiency [3]. To assess the quality of run-in to the CPU by the parameter of fuel consumption stabilization in real time on each mode of running tests, the functioning of the entire technological system is regarded as a random process $F_z(t) = F[Z(t)]$,

where:*Z(t)*- set of states of the elements of the system.

$F_z(t)$ – a set of states of elements functioning in real time.

The mathematical expectation of the fuel consumption stabilization criterion (1) is taken as the indicator of the quality of the functioning at the time I, and for the performance characteristic of the criterion for the stabilization of fuel consumption, the time interval [a, b] is its mathematical expectation with the output effect of the system on this interval:

Instant indicators of the quality and reliability of the system are defined as the ratio of the performance indicators of the functioning of a real system to the corresponding indicators of an ideal system. Interval indicators are defined as the ratio of the output effect of a real system to the corresponding parameters of an ideal system or averaging in the interval [a, b] an instantaneous reliability index.

In work [4] classification characteristics for the purposes of selection of qualitative indicators of energy-saving technologies (ECT), including technologies of reduced running tests, are considered more fully.

In accordance with the "Nomenclature of energy efficiency indicators and the procedure for their introduction into normative documentation" (DSTU 3755-98) and depending on the composition and characteristics of the tasks to be performed, the quality indicators of energy-saving technologies (ECT)

IOP Conf. Series: Journal of Physics: Conf. Series **1066** (2018) 012017 doi:10.1088/1742-6596/1066/1/012017

are classified according to various characteristics, and not only by the uniformity of properties [5 -9]. Universal signs of classification of quality indicators of ECT as applied to the tasks of providing and assessing the quality of ECTs for shortened run tests are:

-energy efficiency factor of KBE,%;

-specific consumption of energy resource b, J / unit of output (work);

- specific fuel consumption bn, (UT in t) / unit of production (work);
- energy intensity of the basic production assets Eof, J / UAH;
- rated voltage UN, V;
- rated current IN, A;
- frequency f, Hz.

The quality of ECT, depending on the number of indicators, can be presented in the form of single, complex and integral parameters. Depending on the form of representation of the characteristic properties, absolute, relative and specific quality indicators of ECT are distinguished.

Of particular importance are the evaluation indicators, characterizing the economy and manufacturability, i.e. those properties that form the quality parameter of ECT as a technological object of the enterprise railway. transport or operated technological object. Estimated indicators are used to normalize the quality of ECT, assess its technical level in the development and standardization, to verify the quality parameters when it is controlled, tested and certified. According to the evaluation indicators, various ECTs classified as classifiers are classified in terms of manufacturability.

Estimates of the quality of ECT are divided into two groups:

• indicators that are used to compare competing technologies in the market by the degree of customer satisfaction with certain economic or environmental parameters;

• indicators that are used to verify compliance with the requirements of international and domestic standards, mandatory for implementation. First of all, this refers to the requirements for safety and environmental friendliness of ECT.

Estimated indicators are also grouped according to the homogeneity of the characteristics being characterized into three types: functional, resource-saving, nature-conservative.

Indicators of functional suitability characterize the technical essence of technology, the properties that determine the basic functions for which it is designed, i.e. the ability of technology (in our case, ECT) to fulfill its own economic or environmental functions in real operating conditions. By appointment, ECT unites such groups of quality indicators as reliability, ergonomics and aesthetics.

The reliability indicators of energy-saving technology characterize its ability to store real-time operating values within specified limits for the values of all given quality indicators, while observing the prescribed modes and conditions of use in maintenance, repair, etc.

Indicators of ergonomic ECT characterize its adaptability to human use in technological processes of railway transport enterprises.

The esthetics of esthetics of ECT characterize its aesthetic impact on humans and are intended for quantitative and qualitative assessment of aesthetic value, the degree of conformity of ECT to the aesthetic demands of certain groups of consumers in specific conditions of railway transport.

Resource-saving quality indicators of ECT characterize its properties, which determine the level of spent material resources in the creation and operation for the formation, provision and realization of its quality.

The group of resource-saving indicators includes subgroups of indicators of manufacturability of ECT and resource consumption.

Indicators of technological effectiveness of ECT characterize the features of its composition and structure, affecting the level of costs of raw materials, materials, fuel, energy, labor and time in operation.

A special group consists of environmental quality indicators of ECT, characterizing its properties associated with the impact of technology on human beings and the environment in the processes of its implementation at enterprises or technical objects (rolling stock) of rail transport. Environmental quality indicators of ECT are combined into two groups of indicators for safety and environmental friendliness.

Classification indicators are applied to the EST of any kind in question - at enterprises and technical objects (rolling stock) of railway transport, as they determine the areas of its application and conditions of use. Classification indicators play an important role in the identification of products.

Energy-saving technologies of railway transport use indicators of functional suitability, maintainability and manufacturability.

The final choice of the nomenclature of quality indicators for the ECT is made taking into account the requirements of international standards ISO 9000-2001, ISO 9004-2001, ISO / TR 10013: 2003, ISO 19011-2003.

The determining factor in the choice of the estimated quality indicators for ECT is the minimization of the costs for the development, implementation and operation of ECT, while simultaneously improving qualitatively the performance of Russian rail transport in a market economy.

When choosing and determining the criterion for assessing the quality of run-in of a cylinder-piston group, various dependencies and qualitative indicators of energy-saving technologies are considered. The possibilities of maintenance and control of the technology of reduced running tests in real time mode, the nature of the main processes determining the transition of technology to the marginal state, the possibility and method of restoring the technical resource, the availability in the technological hierarchy of computers or other devices of discrete technology are taken into account [11].

Yu.B. Germeier formulated the principle of guaranteed result of the quality of technology (technical object), which is that for this economic efficiency criterion, efficiency evaluation should be based on obtaining the guaranteed (maximally guaranteed) value of the efficiency criterion. With regard to the tasks of ensuring the quality of ECT indicators, this principle is interpreted as comparative in the choice of solutions based on guaranteed values of quality and reliability indicators in real-time operation.

When choosing the criterion for stabilizing fuel consumption, which evaluates the quality of the run-in of the cylinder piston group in real time, it is necessary to take into account that the regimes on which the main run-in of the cylinder piston group (GPG) occurs are non-nominal (n = 300 min-1 - n = 450 min-1) , are characterized by low indicators of the work process. The reason for low indicator indicators is considered to be an unstable fuel supply process, characterized by fuel injection skips and uneven injection of fuel through cycles and cylinders. As a result, - in working cylinders on non-nominal modes, power distributions are constantly occurring at the set load of the running mode. Taking into account this factor, the desire to choose the criterion for the quality of running-in of a pair of friction piston ring-cylinder liner of a GPG in real time of running tests becomes clear.

The choice of the criterion should provide a methodology for non-destructive quality control of the running-in of the GPG in real time on all running tests of diesel locomotives on the diesel locomotive frame. The application of the criterion providing a methodology for non-destructive quality control of the running-in of the GPG is the basis for the creation of a reduced energy-saving technology of factory running tests.

## Analysis of previous studies and formulation of the problem

Analysis of the running-in quality of the friction pair "piston ring-cylinder liner" pair of diesel locomotive diesel locomotives on the main, non-nominal running conditions showed that the main influence on run-in has a fuel supply process, characterized by injection skips and uneven injection of fuel in cycles and cylinders. Additionally, the factor of sudden switching-off of auxiliary equipment: compressor, main and additional cooling water cooling circuits of the diesel generator, cooling fans of traction motors, also has an effect.

It is known from literary sources that the factors cited are very negative and their influence (throughout the life cycle) has not been fully studied to the present.

The current practice of assessing the quality of running-in of GPG diesel locomotives in factory and depot conditions is based on methods of destructive testing, which in itself is not only inefficient, but requires large material costs.
A number of researchers used the method of quality control based on selective measurements of fuel consumption. However, in this case, the account of the influence of the on-off of the auxiliary equipment was not applied to the amount of fuel consumption.
As practice has shown, the methods used do not allow to give an accurate estimate of the quality of the running-in of the GPG mirror by the nature of the change in the fuel consumption in real time.

**Results of the conducted studies**

The mathematical model of the criterion for the quality of run-in to the GPG was developed on the basis of works [12-16], which allowed to substantiate the theory of a stable fuel supply process in the entire range of rotation frequencies and loads of diesel locomotives. Further development of the theory is confirmed by a special design nozzle [17], which in practice ensured a stable fuel supply law in all operating modes of diesel locomotives.

A stable fuel supply law implemented by special design injectors allowed us to justify the mathematical modeling of the criterion that assesses the qualitative side of the running in of the diesel engine's CPU, both at the diesel test station and the diesel locomotive frame, in real time.

The software of the mathematical model is implemented in a certified fuel consumption meter AK-DT 0.5, which allows you to record the moment of stabilization of fuel consumption, chosen by us as a criterion for fuel consumption stabilization ($K_g^{est}$), in real time on each run-in mode.

The chosen criterion allowed to provide a method of non-destructive quality control of run-in of the GPG in real time at each run-in test mode.

The chosen criterion, having confirmed its effectiveness at the diesel test station, required additional mathematical modeling and refinement of the software, taking into account the suddenness and frequency of switching on and off auxiliary diesel equipment on the diesel locomotive frame and which had never been considered before.

Modeling on-off of auxiliary equipment and their influence on the criterion ($K_g^{est}$), stabilization of fuel consumption was made on the basis of the law of normal distribution.

At the same time, the mathematical model was optimized taking into account the choice of regimes for short running tests. The software of the developed mathematical model is realized within the framework of non-destructive quality control of run-in of the GPG in real-time tests.

When developing a mathematical model, a special difficulty was created by the apparatus for monitoring and recording the sudden on-time ($t_{rec}$) switching on / off of the main water cooling fan ($t_{cf}$) and the brake compressor ($t_{bc}$).

For this case, the effect on the criterion for stabilizing fuel consumption has the form:

$$Kg^{est} = t_{ex} + \frac{\kappa \cdot r}{r - f_{\text{mod}}}(t_{cf} + t_{bc} + t_{rec}), \quad (1)$$

Where tex - time of running tests; k- the density of the running tests; r is the running time; - part of the operating time of the diesel when the auxiliary equipment is switched on and off suddenly.

Increasing the on-off duration, during useful use, is undesirable since the duration and affects the Kgest.

In the course of the research it was accepted that the law of distribution of the determining parameter (DP) for sudden on-off of various auxiliary equipment is a function of the sudden periodicity of the DP. As the duration of the test increases, the amount of information about the course of the influence of the DP increases, and, consequently, the degree of the random process decreases. To determine the dependence (Kgest) of the resistance of the DP, the sudden on-off delay is approximated by the hyperbolic dependence

$$\sigma_V = \varphi(t_{\text{rec}}) = \frac{1}{p + q_{bc}}, \tag{2}$$

where q is the coefficient characterizing the rate of decrease of the variance

$$p = \frac{1}{\sigma(0)}; \tag{3}$$

In this case, the dependence of $K_g^{est}$ on the on / off time of the auxiliary equipment takes the form:

$$Kge[\varphi(t_{\text{пр}})] = \frac{S_0 - S_{\text{cv}}}{V_2[\varphi(t_{\text{bc}})]}, \tag{4}$$

where $S_0$ and $S_{cv}$ are the initial and critical values of the DP, respectively; Is the upper bound of the rate of change of the DP.

Thus, the problem is reduced to finding the dependence of the sudden on-off of auxiliary equipment with respect to different laws of the density of the separation of the rate of change in the DP.

In the case of a uniform law of the distribution density, the rate of change in the DP is the expectation and variance of the form

$$m_V = \frac{1}{2}(V_1 + V_2) \tag{5}$$

$$\sigma_V = \frac{V_2 - V_1}{2\sqrt{3}} \tag{6}$$

Thus, the problem of changing in the DP.

In the case of a uniform law of the distribution density, the rate of change in the DP is the expectation and variance of the form

$$\tau = \frac{S_0 - S_{\text{cv}}}{\text{m}_\text{V} + \sqrt{3}\dfrac{1}{p + qt_{rec}}} \tag{7}$$

Equality (7) reduces to the form

$$\tau = \Theta \frac{c_1 + b_1 t_{rec}}{c_1' + b_1 t_{rec}} \tag{8}$$

$$c_1 = m_V p;\ \frac{1}{p} = \sigma_V;\ \Theta = \frac{S_0 - S_{cv}}{\mathrm{m_V}},\ b = m_V q,\ c_1' = c_1\sqrt{3}$$

Similarly, in the case of Simpson's law under

$$m_V = \frac{1}{2}(V_1 + V_2); \tag{9}$$

$$\sigma_V = \frac{V_1 + V_2}{2\sqrt{6}}; \tag{10}$$

$$\tau = \Theta \frac{c_1 + b_1 t_{\text{пр}}}{c_1'' + b_{1t} t_{\text{пр}}}, \tag{11}$$

$$c_1'' = c_1 + \sqrt{6}$$

$t_{rec}$= 0 for (11) fair

$$\tau = \Theta \frac{c_1}{c_1'}. \tag{12}$$

The curve (11) is an equilateral hyperbola, related to asymptotes whose center of symmetry is shifted to a point with coordinates $\left(-\frac{c_1}{c_1'}, \Theta\right)$.

According to Simpson's law, an analogous expression is obtained with the coordinates of the center of symmetry $\left(-\frac{c_1}{c_1''}, \Theta\right)$.

$$\tau = \frac{S_0 - S_{\text{кр}}}{\mathrm{m_V}} \cdot \frac{c_1}{c_1'}. \tag{13}$$

In Fig. 1, 2, which shows the dependencies constructed from expressions (8) and (11) with the values of resistance ($Kg^{est}$) and the duration of the sudden ON-OFF of the DP.

An analysis of the nature of the dependence (between the stability of the chosen criterion and the duration of switching-on and off of various equipment) carried out in the framework of the formulas below

$$S_0 - S_{\text{кр}} = 5 \text{ con.un.};\ V_1 = 0.01\frac{\text{con.un}}{\text{h}};\ V_2 = 0.08\frac{\text{con.un}}{\text{h}};$$

showed that with the increase in the duration of the sudden on-off of various equipment with the dependence $\tau = f(t_{rec})$ the stability asymptotically approaches the parameter.

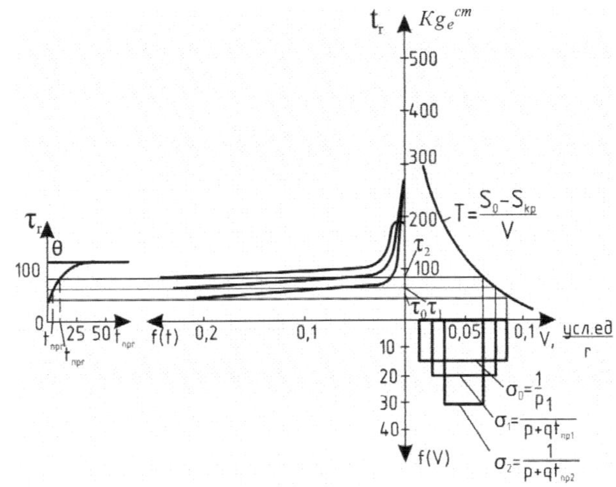

**Figure 1**. The relationship between the resistance ($Kg^{est}$) and the duration of sudden on-off of various equipment at a uniform density of distribution of the rate of change in the DP

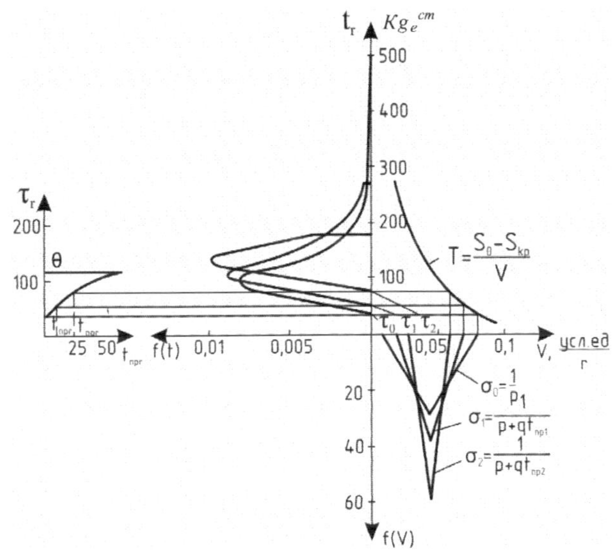

**Figure 2**. Dependence between the resistance ($Kg^{est}$) and the duration of the sudden switching on of different equipment with the density of distribution of the rate of change of the DP by the law of Simpson

In the case of a truncated Weibull distribution [18], which is an asymmetric distribution, a decrease in the variance as a result of the sudden on-off of equipment entails a decrease in the mathematical expectation, so that it is difficult to obtain a precise analytical expression of the dependence of the resistance on the sudden on-off of various equipment. Numerical analysis with various variations of the parameters of the Weibull law shows that the dependence of the resistance ($Kg^{est}$) on the sudden on-off of various equipment can approximately be represented by an empirical relation

$$\tau[\varphi(t_{rec})] \cong \frac{S_0 - S_{cv}}{V_2} + \kappa' t_{rec}, \tag{14}$$

where $\kappa' \cong \frac{a \cdot q}{2}$.

In the distribution of Rayleigh, *a=2,* but with exponential, a=1. Relation (14) is shown in Fig. 3 during $S_0 - S_{cv} = 5$ con.un.; $V = 0.01\frac{\text{con.un}}{\text{h}}$; $V = 0.08\frac{\text{con. un.}}{\text{h}}$; $\Theta = 2$; $q = 2$.

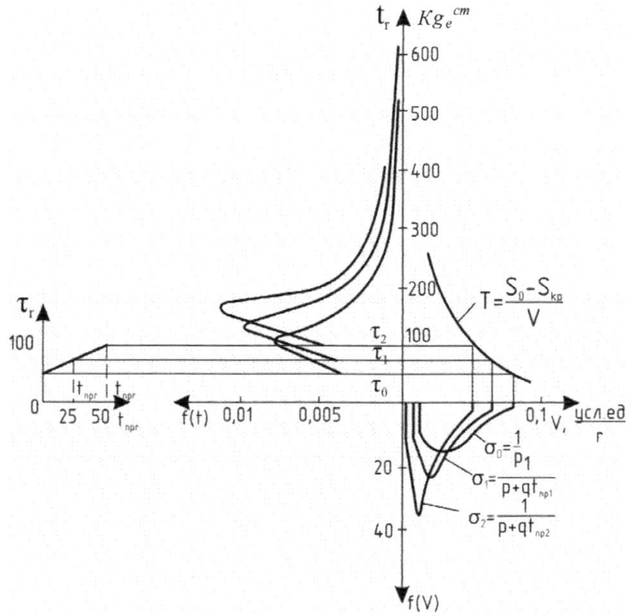


**Figure 3.** Dependence between the resistance ($Kg^{est)}$ and the duration of sudden switching on of different equipment at the rate of change in the rate of change of the DP according to the Weibull law

The dependence $t(t_{rec})$ is a decreasing function from $t_{rec}$ [2] and can be approximated by a function of the form

$$t_B = \frac{1}{c_2 + b_2 t_{rec}}. \tag{15}$$

Using formulas (1), (8), (11), (14) and (15) allows you to get

$$t_и = \Theta\frac{c_1 + b_1 t_{rec}}{c_1' + b_1 t_{rec}} - \frac{\kappa \cdot r}{r - f_{mod}}\left(\frac{1}{c_2 + b_2 t_{rec}} + t_{rec}\right); \tag{16}$$

$$t_и = \Theta\frac{c_1 + b_1 t_{rec}}{c_1'' + b_1 t_{rec}} - \frac{\kappa \cdot r}{r - f_{mod}}\left(\frac{1}{c_2 + b_2 t_{rec}} + t_{rec}\right); \tag{17}$$

$$t_и = \frac{S_0 - S_{cv}}{V_2} - \frac{aq}{2} - \frac{\kappa \cdot r}{r - f_{mod}}\left(\frac{1}{c_2 + b_2 t_{rec}} + t_{rec}\right). (18)$$

The useful life $(t_и)$ has a maximum value $t_{rec} = t_{rec\ opt}$ Value $t_{rec\ opt}$ can be found graphically or as a result of solving equations (16) ÷ (18) to the extremum,

$$\frac{dt_и}{dt_{rec}} = 0. \quad (19)$$

Equations (16), (17) reduce to an algebraic equation of the fourth degree, and equation (18) is of the second power with respect to $t_{rec\ opt.}$

So, values $c_2 = b_2 = 0.04\ \frac{1}{h}$, $\kappa = 1$ $q = 1$, $S_0 - S_{cv} = 5_{con\ un}$; $V = 0.01\frac{con.un}{h}$; $V_2 = \frac{con.un}{h}$ It can be determined that for cases (17) и (18) $\tau_{opt}$ are respectively equal to 7,0 hand 8,2 h. reduced running tests.

It should be noted that in the proposed mathematical model, the criterion for stabilizing fuel consumption ($Kg^{est}$) took into account the optimal volume of technological operations, including both the existing methodology and the methodology for reducing short run tests [19].

**Acknowledgments**
The article was prepared within development program of the Flagship Regional University on the basis of Belgorod State Technological University named after V.G. Shukhov.

**References**

[1] Zonov V D 2005 Mathematical design and research of power descriptions of toplivopodachi in the sprayers of the special kostrukcii (Mezhvuz. sb. nauchn. labours. Khar'kov:ukRGAZht)pp 112-122

[2] Zonov V D, Dan'ko N I 2006 Criterion of control of index of quality of power descriptions by the toplivopodachi sprayers of the special construction of diesel engines Naukovo-praktichniy J. 1 pp 25-28

[3] Zonov V D, Rozenblit G B Grigor'ev and A L 2001 Matematic hnemodely uvannyaprocesupalivopodach at sprayers of special konstrukcZb.nauk.prac/ KhaRDAZt 45 pp 82-88

[4] Zonov V D,Rozenblit G B, Grigor'ev A L 1999 Analysis of reasons and determination of terms, providing impermeability of flat differential valve of forsunki (Dvigateli internal combustion: Announcer KHGPU) pp 82-91

[5] Zonov V D 2007 Prognozno-matematiheskie methods of estimation of quality of energysaving technologies of earning extra money of cilitndro-piston group of diesel engine diesel of Vsnikakadem Ukrani 3-4pp 124-129

[6] Sprayer of specalno konstrukc for a diesel that gazodizel'nogoprocesu model 30267 Ukraïna 7F02Ì45/00( Zayavl 27.07.2007; Opubl.25.02.2008) Byul ¹4

[7] Process of the garyacho rolling dvigunavnutrshnogozgorannya Patent on a korisnumodel 47003( Ukrainaf02b79/00 G01m15/04;04.08.2009; Opubl.11.01.2010) Byul.1

IOP Conf. Series: Journal of Physics: Conf. Series **1066** (2018) 012018 doi:10.1088/1742-6596/1066/1/012018

# METHODS AND DEVICES OF HEAT WASTED GASES UTILIZATION FROM STATIONARY ENGINES OF INTERNAL COMBUSTION

**A V Gubarev , M A Golovkov , D S Dyachuk and S A Bychikhin**

Department of Heat Engineering Technology, Belgorod State Technological University of V.G. Shukhov, Belgorod, 308012, RU

**Abstract**. The article substantiates the urgency of using cogeneration units of low power for supplying energy to objects that can not receive centralized power supply. Various ways and methods of utilizing the heat of exhaust gases of internal combustion engines, as well as the features, advantages and disadvantages of various types of heat recovery agents.

In Russia, as in the main successor of the USSR, the share of centralization of energy supply for consumers of various purposes is very high. With all its advantages, centralized energy supply systems have a number of shortcomings. The most important of these is the availability of expensive and periodically failing energy transport systems (heat and electricity networks). It should be noted that energy losses in transportation systems constitute a significant share of all losses in the power system [1]. Also, significant deficiencies of centralized systems are their low maneuverability, the complexity of centralized regulation of energy supply, and the low efficiency of equipment of such systems and systems as a whole [2]. These shortcomings of centralized systems determine the current trend towards an increase in the share of decentralized systems of low power [3].

It should also be borne in mind that in the Russian Federation the share of districts is large, centralized energy supply and fuel-supply of which is very difficult. First of all, this applies to small settlements and research settlements located far from areas with developed infrastructure.

The solution to the problem of power supply for such facilities is the use of local alternative or renewable sources of energy (if any) or the organization of work of a power source on those traditional fuels whose delivery to these facilities is the least expensive. As such fuels, for example, liquid hydrocarbon fuels (gasoline, diesel fuel) or liquefied gases can be considered. In this case, cogeneration facilities are the most effective as objects generating heat and electric energy. The cost of generating electric and heat energy in such facilities is low, they can be located in close proximity to the consumer, which allows to abandon energy transport systems of a large extent, and also facilitates the regulation of energy release [4].

The use of cogeneration plants as reserve sources of energy supply for heat consumers of the first category and for electricity consumers of the first and second categories is also topical.

The main elements of the cogeneration plant are the primary engine, the electric generator and the heat recovery system. As a primary engine, piston internal combustion engines are very efficiently and reliably used [5].

The exhaust gases of these engines have a very high potential: their temperature, depending on the type of engine, mode of operation and the type of fuel burnt, can range from 350 to 800 ° C. The use of heat from exhausted gases can improve the efficiency of the energy system, reduce its negative impact on the environment, and, in some cases, approach the creation of low-waste technology.

Consideration, analysis and comparison of different ways, methods and apparatus for using the heat of exhaust gases of internal combustion engines to select the most relevant, functional and cheap option for a particular mini-power plant with such engines and the conditions of its operation is the purpose of this work.

There are two main ways to use the thermal potential of exhaust gases of internal combustion engines:

IOP Conf. Series: Journal of Physics: Conf. Series **1066** (2018) 012018 doi:10.1088/1742-6596/1066/1/012018

1) covering the own needs of the power unit;
2) supply of thermal energy to external consumers.

In the first case, due to the use of the thermal energy of the exhaust gases with the help of various devices, for example, a steam and steam installation, a useful work is done, which is either directly transferred to the crankshaft of the internal combustion engine, thereby increasing its power and efficiency, water of various units and engine systems, for example, superchargers of the cooling system. Due to the above options for using the heat of exhaust gases, engine power can be increased by 15%, and its efficiency is increased by 5 ... 18% [6].

However, in the context of the issues discussed in this article, the second way is more important, that is, the utilization of the heat of the exhaust gases for the preparation of the coolant, which is subsequently channeled to the needs of heat supply to external consumers in order to cover the seasonal load and for hot water supply.

As heat-recovery devices in cogeneration plants, the greatest use has been made of water-tube and gas-tube boilers, utilizes and gas water heaters, which can operate in a very wide range of temperatures of engine exhaust gases. The advantages and disadvantages, as well as the scope of application of these facilities, are determined by their design features.

Thus, the advantages of water-tube waste heat boilers, in which the combustion products move in the shell space, and the heating medium in the pipes, is a sufficiently high intensity of heat transfer from both coolants and relatively small values of gas-dynamic resistance. However, in such units there is no possibility of mechanical cleaning of heating surfaces from the gas side, which significantly reduces the scope of their application. That is, in the primary engines of cogeneration units with water-tube boilers, only clean fuels that are not prone to carbon formation can be burned.

When burning fuels prone to soot formation in the engine cylinders, it is more preferable to use gas-tube waste heat boilers in which gases are sent to the pipe space. In this case, the cleaning of the heating surfaces from the gas side is possible both mechanically and chemically, and in a space that is difficult to access the annular space, a much cleaner coolant flows through the water treatment apparatus. Cleaning of the inner tube space from the scale can very effectively be effected by chemical methods. It should be noted that such heat recovery boilers are characterized by a low heat transfer coefficient on the part of the heated coolant, due to the low speeds of its movement. In some areas, coolant stagnation and intensive scale deposition on the outer surface of the pipes may occur, resulting in the burning of such pipes at high temperatures and, consequently, the emergency shutdown of the heat recovery boiler. In connection with the foregoing, the construction of such boilers is somewhat complicated due to the presence of elements contributing to the increase in the speed of the heated coolant, especially in the high-temperature zone of gases, for example, partitions, which allow the multi-way movement of water in the annular space. Another disadvantage of gas-tube heat recovery boilers is the high gas flow losses due to very high values of gas in the pipes, which can cause the need for the smoke generator to be included in the equipment of the cogeneration plant.

A very interesting system for utilizing the heat of exhaust gases of diesel internal combustion engines, which makes it possible to increase the energy efficiency and environmental safety of a mini power plant, was proposed by the staff of the Lugansk Machine-Building Institute [7]. According to the proposed solution, in a system for utilizing the heat of exhaust gases of an internal combustion engine at a power plant including several internal combustion engines and heat recovery devices, at least one of the heat exchangers is performed as a catalytic reactor for converting a hydrocarbon fuel to a hydrogen-containing gas by thermo chemical regeneration of the heat of exhaust gases of one from the engines. In this case, the heat of the exhausted gases is more fully utilized, the hydrocarbon fuels of oil production are saved and the toxic gas emissions to the environment are reduced. As shortcomings of the proposed solution, it is possible to isolate the impossibility of using such a scheme in a cogeneration system that includes one engine, as well as increased safety requirements for the operation of such a facility, since hydrogen has very wide explosive limits in mixture with air - from 4 to 75%.

It is also very promising to use, as heat recovery heat generators for internal combustion engines, condensation water heaters, which not only allow deep cooling of combustion products, but also their wet cleaning, in particular, from soot particles. As a complication arising from the use of such heat recovery devices, one can single out the problem of water contamination with components contained in the exhaust gases of engines. This problem necessitates the use of intermediate heat carriers and additional heat exchange surfaces, which complicates and makes the more expensive construction of the water heater.

Thus, when choosing the design of a heat recovery boiler or a water heater for a cogeneration plant, it is necessary to thoroughly evaluate the conditions of its operation, parameters and characteristics of heat carriers, as well as to make thermal engineering and technical and economic calculations that will allow selecting the most inexpensive and efficient method utilization of heat of exhaust gases of the internal combustion engine and its design.

**Acknowledgments**
The article was prepared within development program of the Flagship Regional University on the basis of Belgorod State Technological University named after V.G. Shukhov.

**References**

[1] Kozhevnikov V and Kuznetsov V 2014 Res. J. of Pharm., Biol. and Chem. Sci. 5 pp 1619–1624
[2] Kuleshov M and Gubarev A 2015 Inter. J. of App. Engin. Res. 24 pp 44872–44878.
[3] Grinenko G, Kozhevnikov V, Kuleshov M, Pogonin A 2012 Prospects for the development of the market of condensing boilers in Russia Bulletin of the Belgorod State Technological University V.G. Shukhov vol 3 pp 145-149
[4] Vedruchenko V, Krainov V, Zhdanov N 2015 Choice of scheme for utilization of heat from exhaust gases of piston ICEs and estimation of useful heat utilization in the composition of a cogeneration plant Omsk Scientific Bulletin vol 1 pp 114-119
[5] Gavrilov K 2017 Stationary and mobile power plants with internal combustion engines (Moscow: FGBU UMC APK of the Ministry of Agriculture of the Russian Federation) p 320
[6] Kovalenko Yu 2003 Improving the efficiency of internal combustion engines by utilizing the heat of their exhaust gases: Dis kand Techn Sciences (Chelyabinsk: ChVAI) p 203
[7] Zvonov V Chernykh V Baranov V 1995 (RU 2046979 Exhaust gas heat recovery system of internal combustion engine)

IOP Conf. Series: Journal of Physics: Conf. Series **1066** (2018) 012019 doi:10.1088/1742-6596/1066/1/012019

# OPTIMIZATION OF COEFFICIENTS SPECIFIED WITH NON-SINUSOIDAL AND NON-SYMMETRY FEEDING VOLTAGE IN ELECTRICAL SUPPLY SYSTEMS IN INDIVIDUAL RESIDENTIAL SECTOR

**M A Averbukh and E V Zhilin**

Department of Power Engineering and Automation, Belgorod State Technological University of V.G. Shukhov, Belgorod, 308012, RU

**Abstract.** In this paper, we consider coefficients that characterize nonsinusoidality and asymmetry. It is revealed that due to the presence of single-phase electric receivers with nonlinear volt-ampere characteristic in the power supply systems of Individual residential sector, their values exceed the maximum permissible values, this leads to the emergence of additional energy losses that amount to 14% of the total power consumption. Therefore, the goal of the work is to optimize the coefficients, according to the criterion of minimum energy losses with a minimum of costs for the installation of technical equipment.

The main consumers of the electricity supply system individual residential sector(IRS) are single-phase household electrical receivers. They have a nonlinear current-voltage characteristic, since they contain semiconductor elements in their structure. This leads to the appearance of higher harmonics of current and voltage in the system of electricity supply of individual residential sector S. In addition, the uneven distribution of single-phase electric receivers in a three-phase four-wire network causes the appearance of an unbalance current in the zero conductor. The presence of higher harmonics and asymmetric connection of electric receivers leads to a decrease in the quality of electricity (QE) in the electricity supply system individual residential sector [1].

To the indicators characterizing the non-sinusoidal operating mode in the systems of power supply individual residential sector you can include the total coefficient of harmonic components of voltage and current [2]:

$$K_U = \frac{\sqrt{\sum_{n=2}^{40} U_{(n)}^2}}{U_1} \cdot 100; \; K_I = \frac{\sqrt{\sum_{n=2}^{40} I_{(n)}^2}}{I_1} \cdot 100.$$

individual residential sector Here $n$ is the ordinal number of the harmonic component; $U\ (n)$ is the actual value of the $n^{th}$ harmonic component.

Thus, in order to estimate the coefficients that determine the non-sinusoidality of periodic curves, it is necessary to know the spectral composition of non-sinusoidal currents and voltages.

Another characteristic of non-sinusoidality is the coefficient of the nth harmonic component of voltage and current:

$$K_{U(n)} = \frac{U_{(n)}}{U_1} \cdot 100; \; K_{I(n)} = \frac{I_{(n)}}{I_1} \cdot 100.$$

The asymmetric mode of operation characterizes the unbalance coefficient of the voltage in the reverse and zero sequence:

$$K_{2(u)} = \frac{U_{2(1)}}{U_{1(1)}} \cdot 100\%; \; K_{0(u)} = \frac{\sqrt{3}U_{0(1)}}{U_{1(1)}} \cdot 100\%.$$

The values of the unbalance coefficients of the stresses in the negative sequence K2U and the unbalance of the voltages along the zero sequence K0U at the point of transmission of electrical energy averaged over a time interval of 10 minutes should not exceed 2% in the course of 95% of the time and 4% % of the time interval in one week.

The high level of non-sinusoidal and non-symmetric coefficients in the power supply systems of the private housing construction leads to the appearance of additional energy losses, which can reach 14% of the total consumed electricity [3].

To reduce the coefficients, and as a result of additional losses of electricity, there are many different methods and technical devices [4]. However, this will require additional financial investments. As a result, there are two interrelated components of costs: the cost of installing technical devices to reduce the coefficients and costs from poor quality of electricity.

Applying various methods that reduce the coefficients of nonsense and non-symmetry in the electricity system of IRS is very difficult, and sometimes impossible, and the use of various technical devices does not always pay off even for the entire period of operation.

Therefore, the main purpose of optimizing the coefficients characterizing the non-sinusoidality and asymmetry is the choice of technical means to compensate for excess reactive power, distortion current, and unbalance current from an asymmetric load. The optimization problem is solved by minimizing the energy losses caused by the presence of higher harmonics and unbalanced currents.

Setting optimization task is to determine the object, parameters and optimization criteria, as well as restrictions for the objective function.

The choice of the optimal type, capacity and location of technical means to reduce power losses in the system of power supply individual residential sector is performed taking into account two criteria: a minimum of the resulted costs and a minimum of losses of active capacity. The objective function for the first criterion$Z(Э)$ contains two terms: the cost of electricity losses in the electricity supply system of the housing construction company, the cost of installing technical equipment.

$$Z(\Delta Э) = c_1 \cdot \Delta Э + K_H \cdot K(Э) \rightarrow \min,$$

where, c1 - cost of losses of 1 kW • h of electricity; ΔE - electric power losses for a specified period; KN - normative factor of recoupment of investments; Z (Э) - capital costs for the installation of technical equipment.

The target function is subject to the restriction area, which is conditioned by the technical requirements for the individual residential sector power supply system.

$$\begin{cases} |U_1(Y_{PU})| \le U_{(1)EX}; \\ K_U(Y_{PU}) \le K_{U\,EX}; \\ K_I(Y_{PU}) \le K_{I\,EX}; \\ K_{2U}(Y_{PU}) \le K_{2U\,EX}; \\ K_{0U}(Y_{PU}) \le K_{0U\,EX}; \\ \cos\varphi_{A,B,C}(Y_{PU}) \le \cos\varphi_{A,B,C\,nor}, \end{cases}$$

Where$Y_{PU}$ is the vector of the conduction column of the PU, Sm; $U_{(1)ex}$— current (allowable) direct sequence voltage in the network node, V; $cos\varphi_{A,B,Cnor}$ - current (normative) value of the power factor; $K_{2U}$- current (permissible) value of the voltage unbalance factor in the reverse order,%; $K_{0U}$ - current (permissible) zero-sequence voltage unbalance coefficient,%; $K_U$ - current (permissible) value of the total coefficient of harmonic components by voltage,%; $K_I$ -current (permissible) value of the total coefficient of harmonic current components,%;

The system of power supply IRS is described by the method of nodal potentials. The initial data are technical and economic indicators, as well as the cost of electricity, unit costs per unit of technical means, network parameters, load range[5].

$$\dot{I} = \underline{Y} \cdot (S^t((S\underline{Y}S^t)^{-1} \cdot S(\dot{J} - \underline{Y}\dot{E})) + \dot{E}) - \dot{J},$$

where $E$- is the column vector of complex branch electromotive force (non-zero values for the branches of the transformer), B; $J$- is a column-vector of complex node currents (non-zero), A; $Y$ is the diagonal matrix of complex conductivities of branches, Cm; $I$- is the column vector of complex branch currents, A; $S$- is the coupling matrix.

The result of the calculations is the optimal vector of imaginary components of the conductance of branches with a minimum level of higher harmonics and asymmetry, on the basis of which the numbers of branches containing technical devices for compensation of nonlinearity and asymmetry are determined. Terms of realizability of technical means are determined by the accuracy of compensation, speed, operational reliability, unit costs for installation and operation. Therefore, it is necessary to analyze all methods of compensation for higher harmonics and asymmetrical currents in the system of electricity supply of IRS, taking into account its features.

For example, consider a fragment of the power supply system IRS, shown in Fig. 1. For the nodes of the power supply system, it is required to determine the value of the third harmonic compensation current, based on the condition of minimum total costs for installing these devices and covering the losses of active power.

The initial data is the mains voltage ($U_{nom}$= 0,4 kV), the resistance of the line ($R_1$ = 0,4; $R_2$ = 0,5), the third harmonic current at the nodes ($I_{1(3)}$ = 3,12 ; $I_{2(3)}$ = 3,62 ), unit costs for the installation of technical devices ($Z_0$ = 0,5 y.e./A), ), unit costs for covering the losses of active power ($C_0$ = 10 y.e./ kW).

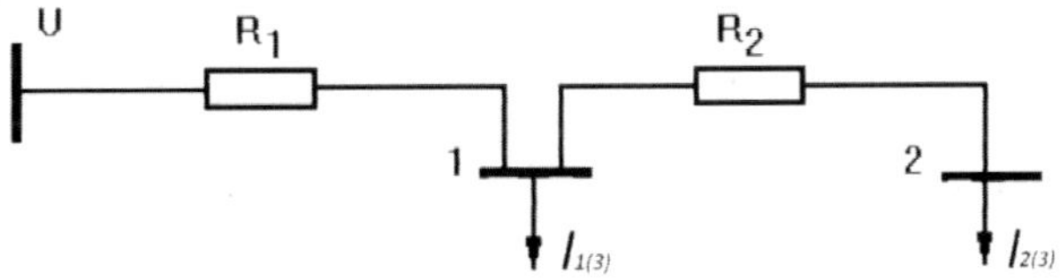


**Figure 1**. Fragment of the electricity supply system IRS

The objective function is the total cost of installing technical devices and covering the losses of active power in the power supply system IRS:

$$Z = Z_0 \cdot (I_{\kappa 1} + I_{\kappa 2})^2 + a_1 \cdot (I_{1(3)} + I_{2(3)} - I_{\kappa 1} - I_{\kappa 2})^2 + a_2 \cdot (I_{2(3)} - I_{\kappa 2})^2 \to \min ,$$

where$a_1 = R_1 \cdot C_0 \cdot 10^{-3}$; $a_2 = R_2 \cdot C_0 \cdot 10^{-3}$, $I_{\kappa 1}$и$I_{\kappa 2}$ – compensation current.

Introduction of a numerical coefficient $10^{-3}$ It is necessary to reduce all components of the objective function to one dimension (standard units).

To solve the problem, we choose the coordinate-wise "descent" method. For this, it is necessary to determine the partial derivatives of the objective function with respect to the variables$I_{\kappa 1}$and$I_{\kappa 2}$.

$$\frac{\partial Z}{\partial I_{\kappa 1}} = Z_0 - 2 \cdot a_1 \cdot (I_{1(3)} + I_{2(3)} - I_{\kappa 1} - I_{\kappa 2});$$

$$\frac{\partial Z}{\partial I_{\kappa 2}} = Z_0 - 2 \cdot a_1 \cdot (I_{1(3)} + I_{2(3)} - I_{\kappa 1} - I_{\kappa 2}) - 2a_2 \cdot (I_{2(3)} - I_{\kappa 2}).$$

We take the initial approximation $I_{\kappa 1}$ = 0, $I_{\kappa 2}$ = 0. For these values, we calculate the values of the objective function and its partial derivatives: $Z^0$ = 22,8 standart units, $\partial Z/\partial I_{\kappa 1}$ = 0,45, $\partial Z/\partial I_{\kappa 2}$ = 0,41.

Obviously, in the direction of the variable$I_{\kappa 2}$ the target function Z is stronger than in the direction $I_{\kappa 1}$, since the derivative $|\partial Z/\partial I_{\kappa 1}| > |\partial Z/\partial I_{\kappa 2}|$.

Therefore, in the direction $I_{\kappa 2}$ we begin the "descent". We take the step size λ = 4 A, then the first approximation will be $I_{\kappa 1}$ = 0, $I_{\kappa 2}$ = 4, and the value of the objective function will be $Z^1$ = 22,725 cu. The second step is $I_{k1}$ = 0, $I_{k2}$ = 8, the value of the objective function $Z^2$ = 22.81 standart units.

From the results of the calculation of the objective function, it is clear that the "descent" along the coordinate $I_{k2}$is worthwhile to cease, because $Z^2 > Z^1$, and return to the values of the variables obtained in the first step. Performing a new second step in the direction of the first variable $I_{k1}$ = 4, $I_{k2}$= 4, the value of the objective function $Z^3$ = 22.79 cu. Movement in the direction of the first variable is inadvisable because $Z^3 > Z^2$.

The point with the coordinates $I_{k1}$ = 0, $I_{k2}$= 4 is in the vicinity of the minimum of the objective function Z, with the adopted step length λ = 4 A, a more accurate solution can not be obtained without the use of computational programs. Thus, for the most effective reduction of the coefficients, it is necessary to install a filter compensating device in the second node with a compensation current of the third harmonic of 4 A.

In conclusion it should be sail thatthe analysis of the power supply system IRS, showed the presence of higher harmonics and currents of asymmetry, which lead to additional losses and an increase in the coefficients characterizing non- sinusoidality and asymmetry.

The choice of technical devices to reduce the coefficients characterizing non-sinusoidality and asymmetry must be performed simultaneously by several parameters, by solving a nonlinear optimization problem.

**Acknowledgments**
The article was prepared within development program of the Flagship Regional University on the basis of Belgorod State Technological University named after V.G. Shukhov.

**References**
[1] Vinogradov A and Zyabkina O 2013 The indicators of the quality of electric energy, caused by the use of LED lamps *Bulletin of the Belgorod State Technological University. V.G. Shukhov* vol 1 pp 159-161
[2] GOST 32144-2013 *Electric Energy Compatibility of technical means is electromagnetic* Norms of quality of electric energy in general-purpose power supply systems
[3] Averbukh M and Zhilin E 2016 On the loss of electricity in the systems of power supply for individual housing construction *Energetik* vol 6 pp 54-57
[4] Averbukh M, Kuznetsov V, Korzhov D 2013 Problems of ensuring electromagnetic compatibility in elecrical installations of industrial enterprises*Bulletin of the Belgorod State Technological University named after VG Shukhov* 2013 vol 5 chapter 36 pp 203-208
[5] Tretyakov E A 2011 Optimization of quality and losses of electric energy in electric networks of non-tensional consumers Transport of the Russian Federation *J of Science, Practice, Economics* **3** (34) pp 50-54

# PARTICULAR QUALITIES OF EXECUTION OF THE PROGRAM COMPONENT FOR ENERGY RESOURCES MANAGEMENT OF THE BELGOROD REGION

**P A Trubaev[1], A V Bulanin[2], K J Shirrime[1] and Yu A Koshlich[3]**

[1]Department of Heat Engineering Technology, Belgorod State TechnologicalUniversity of V.G. Shukhov, Belgorod, 308012, RU
[2] Budget institution Energy Saving Center of the Belgorod region, Belgorod, 308007, RU
[3] Department of Power Engineering and Automation, Belgorod State TechnologicalUniversity of V.G. Shukhov, Belgorod, 308012, RU

**Abstract**. The article considers modern operating systems for increasing energy efficiency at the municipal and regional levels. Present successfully applied practices in other regions are given. The essential distinctive features of the proposed system for the Belgorod region are revealed. Described, developed methods and algorithms for managing energy conservation in the budget sector. The structure of complex interaction of the developed modules used in the created software component is offered.

The system of urban energy resources management is a set of strategic and practical measures aimed at achieving sustainable and efficient consumption of fuel and energy resources and includes both organizational and technical measures at sites located in the city management [1-3]. Currently, several regional systems are operating or being developed in Russia, the review of which is given in Table1.

Based on the analysis of advantages and disadvantages in existing software systems and systems, as well as taking into account the needs of regional and municipal authorities, the following structure of the regional information and analytical system Energy Resources Management Systems of the Belgorod Region "EnergyRegion"(RIAS SUER"EnergyRegion") is proposed.

The goal of the development is to make an operating system that helps manage energy saving and save on the basis of data collection, analysis, problem identification and effective control over their solution. The system should: 1) monitor the efficiency of allocating budget funds to pay for communal resources; 2) to choose the directions of investing in energy saving from the point of view of maximum savings; 3) be the basis for the development of energy service contracts and monitor their implementation.

Main functions of the system are defined:

– storage and analysis of data on energy consumption, identification of falsification of meter readings;
– monitoring the quality of the supplied resources in the building, not depending on the date of collection of energy consumption data;
– calculation of individual energy consumption standards for buildings;
– monitoring of changes in consumption of objects, forecasting of possible emergency situations;
– monitoring and setting of targets for resource savings and monetary costs (limits) for each building;
– forecasting the consumption of energy resources for a long period, setting limits on the consumption of energy resources, monitoring their implementation;
– calculation of actual specific indicators of consumption of energy resources;
– control of energy consumption indicators;
– assessment of energy saving potential for standard energy-saving measures;
– evaluation of the results of the implementation of energy-saving measures;
– formation of administrative documents and control over its execution;
– formation of reporting documentation on the actual implementation of energy service contracts.
– formation of energy efficiency rating.

IOP Conf. Series: Journal of Physics: Conf. Series **1066** (2018) 012020 doi:10.1088/1742-6596/1066/1/012020

**Table 1.Regional and municipal energy management systems [1, 2, 4, 5]**

| | |
|---|---|
| GIS Energy Efficiency | State information system in the field of energy saving and improving of energy efficiency.<br>Disadvantages:<br>1. These are systems of statistical observation that are being organized -as a base of materials in the field of energy saving and a catalog of branch organizations, there are no analytical tools.<br>2. The statistical period is the year.<br>3. "Manual" data collection.<br>4. "GIS" Energy Efficiency "- a temporary and non-systemic solution" (Igor Kozhukhovsky, deputy director of REA, http://www.energy2020.ru/news/news10237.php). |
| Information-analytic system of the fuel and energy balances - IAS TEB (Ivanovo State Energy University) | Implemented:<br>• Orenburg region;<br>• Ryazan Oblast;<br>• Vladimir region;<br>• Republic of Tatarstan;<br>• Ivanovo region.<br>Disadvantages:<br>1. Focuses on enterprises that supply energy, not consumers.<br>2. The analyzed period is a year.<br>3. Technical features of objects (number of stores, availability of insulation, etc.) are not taken into account.<br>4. "Manual" datacollection. |
| City Energy Management System - SGUER (UNDP / GEF Project 00074315 "Energy Efficiency of Buildings in North-West Russia", 2010-2015) | It is developed on the basis of EMIS platform.<br>Implemented by:<br>• Pskov region.<br>• Arhangelsk region.<br>• Vologda Region.<br>Disadvantages:<br>1. It is at the stage of concept development and design decisions.<br>2. The analyzed period is a year.<br>3. Technical features of objects (number of stores, availability of insulation, etc.) are not taken into account.<br>4. "Manual" datacollection. |
| Energy Management Information System - EMIS (Croatia) | Implemented by:<br>• Croatia (official energy management system).<br>• Slovakia (official energy management system).<br>• Pskov Region (a pilot project of 485 projects under the UNDP / GEF program).<br>• Vologda region (pilot project of 16 objects).<br>Disadvantages:<br>1. The system was created to control the payment of bills for energy.<br>2. Does not contain any instrumental analysis - no ranking, grouping, forecasting.<br>3. Data are not provided for comparable conditions.<br>4. The user performs the entire analysis manually, using data upload in Excel format.<br>5. Not adapted to the conditions of Russian office management and administrative management. |
| Information system energy saving LLC Ansists Systems of dispatching, monitoring, AMR, ASUE, AIIS, etc. | Regional GIS Segment Energy Efficiency.<br>Under development.<br>Disadvantages:<br>The system, like the GIS Energy Efficiency, is designed to collect information and maintain documentation with the addition of functionality to record energy service contracts and contracts with suppliers of fuel and energy resources.<br>Are superstructures over the technical means of instrument accounting.<br>There are numerous number of such systems, their main drawback is only accounting and control, without analyzing and evaluating the efficiency of energy consumption. |

As part of this project (RIAS SUER"EnergyRegion") provides for the establishment of monitoring and analysis of energy efficiency of budget consumers of energy resources that are connected to the unified automated information-measuring system of energy accounting (AIS), created in the framework of the regional project number 10081730 "Creation of a regional automated unified automated information - measuring system of energy resources accounting AIS ".

The structural diagram of the system is shown in Fig. 1.

The input data is divided into three groups:

– monitoring of energy consumption: actual and planned consumption of energy resources and water;

– technical data: technical and functional characteristics of facilities, their equipping with accounting devices;

– energy saving indicators: energy efficiency programs for institutions, data on energy service contracts and the results of their implementation.

After the final development of the RIASSUER"EnergyRegion" will contain data on all municipal budget institutions and Resident sector in the Belgorod region.

All collected information will be stored in a single data warehouse with the necessary redundancy and ensuring a work rate sufficient for the functioning of the system.

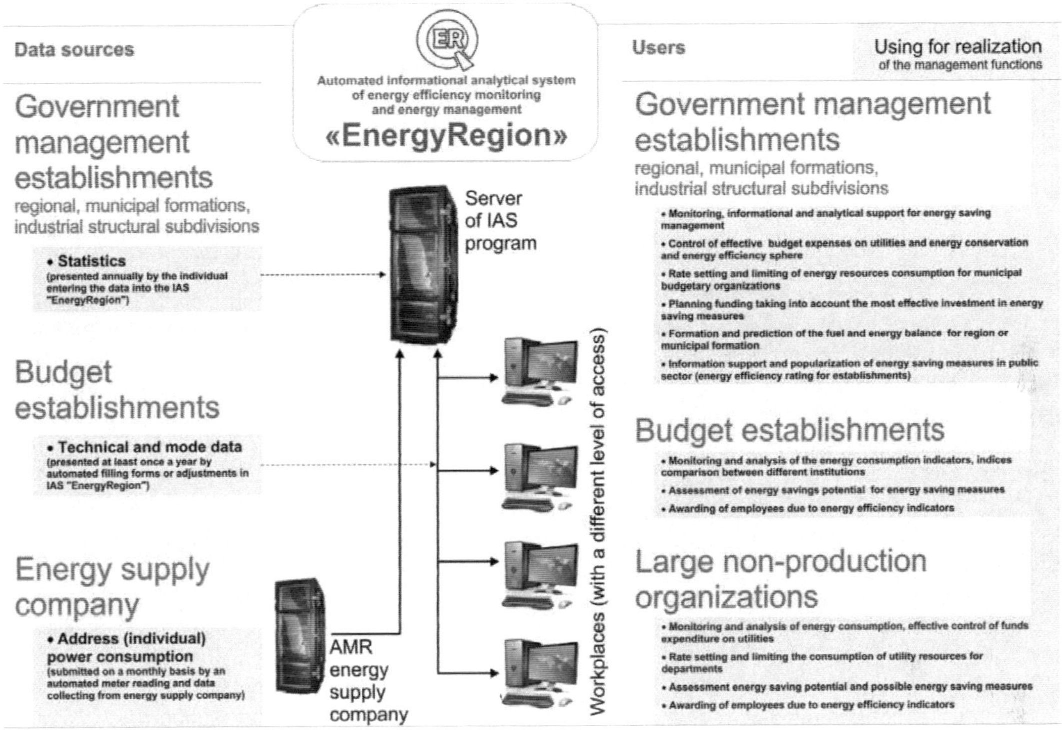


**Figure1.** Schematic representation of the structure of the pilot project of RIAS SUER"EnergyRegion"

To create a system, it is necessary to develop new methods and management tools:

– development of calculation and analytical methods for determining the cost of heating;

– development of a methodology for determining individual energy consumption standards;

– development of a methodology for forecasting energy consumption for a long period;

– development of a methodology for calculating actual specific energy consumption indicators, monitoring energy consumption indicators based on their comparison with norms.

One of the important tasks of the system is the formation and control of limits. The established norms of consumption (limits) are the legislative basis for regional control over budget institutions. Limitations are proposed to be made on the basis of a set of three methods (Fig. 2): a) statistical, according to consumption for the past time for similar objectives, b) analytical and analytical on the basis of technical data objects and modes of operation, c) Acting Building Norms and Rules and codes of regulations.

Thus, the system is aimed at constant monitoring, control of each of the city administration's facilities for the consumption of fuel and energy resources on a systematic basis, taking into account the individual characteristics of each facility. At the same time, the system makes it possible to react quickly to various changes in energy consumption and to evaluate it depending on the time of day, season and weather conditions. New methods and tools for managing energy saving in the budget sector, which will be implemented in the information system will ensure compliance with the requirements of existing regulatory and legal acts in the field of energy saving, timely preparation of up-to-date information on energy saving and consumption of fuel and energy resources and water, as well as its presentation for decision-making by the executive authorities and local self-government of the region.

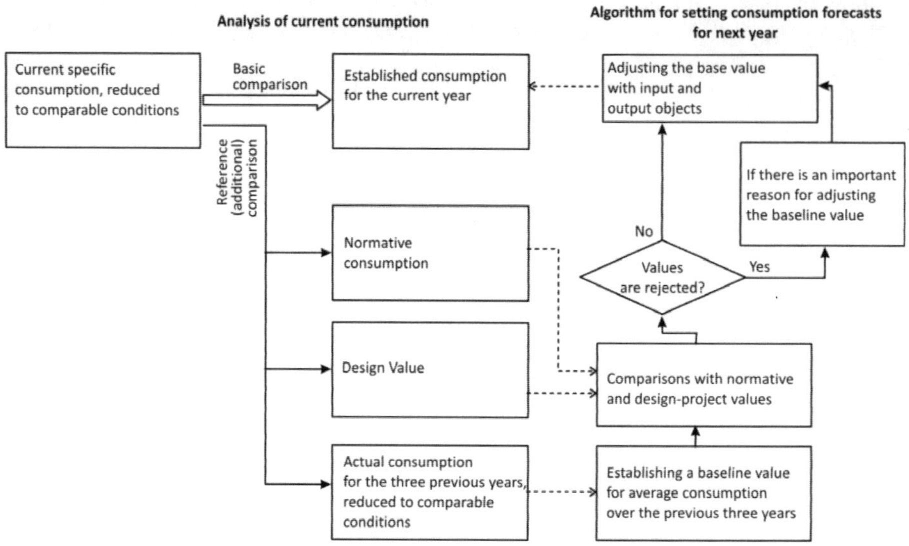


**Figure 2.** Interaction of blocks of analysis and formation of limits

Distinctive features of the proposed system are:

– the possibility of implementing the system in any institution (commercial, state), taking into account the type of activity of the facility;

– integration with existing workflow systems and an operational monthly analysis for management adjustments;

– obtaining the necessary information at a minimum bureaucratic load;

– compilation of a single energy balance of the whole municipal education and evaluation of the most investment-attracting events.

The key advantage of the system is the basing on successful principles of the analysis of energy management systems, where the main task is the analysis of dynamics and rational use of thermal and electric energy in urban facilities. Thus, the complex analytical processing of information within the framework of the RIAS SUER"EnergyRegion" allows us to identify the potential for reducing fuel consumption in municipalities that ensures rational and efficient use of resources, as well as provide support for making informed decisions on the development of the region as a whole.

In conclusion, it should be noted that such technical implementation provided for in the executable software module RIAS SUER"EnergyRegion"enables not only to perform effective operational analysis to respond to various changes in energy consumption and consumption regulation, but also to forecast the most effective measures in further.

## Acknowledgments

The article was prepared within development program of the Flagship Regional University on the basis of Belgorod State Technological University named after V.G. Shukhov.

*The article was prepared within the framework of the Base University Development Program on the basis of the BSTU named after. V.G. Shukhov.

## References

[1] Becker V, Dodonov A, Lukina-Lebedeva M 2016 *Practical Steps to Implement the Urban Management of Energy Resources* (St. Petersburg Method. Instructions) p 49

[2] Ratmanova I, Zheleznyak N 2006 Approach to the organization of infor-mation support of the state policy in the sphere of management of the fuel complexes in the region *Vestnik ISEU* vol 4

[3] Kushchev L and Dronova G 2008 Ways to reduce energy costs in the housing and communal services *Bulletin of Belgorod State Technological University V G Shukhov* vol 2 pp 24-25

[4] Shirrime K and Trubaev P 2016 Analysis problems in the implementation of regional energy management systems *Energy, control and information systems* (Publishing House of BSTU) pp 383-386

[5] Hooke J 2004 Energy Management Information Systems: Achieving Improved Energy Efficiency: *A Handbook for Managers, Engineers and Operational Staff Environment Canada* p 93

# PROBLEM TASKS IN THE FIELD OF RESOURCE AND ENERGY SAVING TECHNOLOGIE

**L A Sivachenko[1], V S Sevostyanov[2] and T N Il'ina[3]**

[1]Department of Transport and Technological Machines, Belarusian-Russian University, Mogilev, 212000,Belarus
Department of Technological Complexes of Machines and Mechanisms, Belgorod State Technological University of V.G. Shukhov, Belgorod, 308012, RU
[3]Department of heat and gas supply and ventilation, Belgorod State Technological University of V.G. Shukhov, Belgorod, 308012, RU

**Abstract**. The current problems of the rational use of mineral raw materials, including the stages of extraction, processing and utilization of wastes using modern technologies are considered. Classification is proposed and a list of the main directions of resource and energy saving in the production environment and in the sphere of related processes and factors is given.

Of all the minerals extracted, which is about 150 billion tons per year, only about 2% are used as useful products, while the rest in the chemically small-modified state is waste [1].
If to add to this huge volumes of processing and the corresponding amounts of waste in agriculture, food industry, chemical production, timber processing, household sector and many other sectors, the problem of resource-saving becomes world-wide.

More than 90% of all production costs ultimately connected with the wasteful use of resources of all types are carried out at large industrial facilities [2]. Therefore, in order to implementthe problems of energy saving, an urgent task is to search for and assess sources of reducing resource consumption in basic sectors of the national economy, mainly in the processes of processing and transformation of various materials and obtaining finished products

The purpose of work:

- identify the main areas of increasing the efficiency of the technological production sphere;
- highlight the most important components that have a significant impact on the rational use of all types of resources;
- to present ways of solving the problems of energy and resource saving in the technologies of various sectors of the national economy.

For a holistic perception of the analyzed problem, the entire system of technological resource saving will be conditionally divided into the production environment and the sphere of adjacent processes and factors. We will consider each of its componentsmore specifically.

Production environment of technological resource-saving:

- optimization of exploration and production of minerals;
- reduction of fuel consumption and primary energy;
- primary preparation and processing of raw materials;
- increasing the degree of processing and extraction of mineral products
- increaseof materials and waste in the share involvement in the production of raw materials,;
- synthesis of new compounds and compositions;
- optimization of technological processes and increase of their non-waste technology;
- Increase the service life of the products.

It is appropriate to initially allocate exploration and extraction of minerals, as a huge amount of materials must be produced and the largest volumes of primary waste are formed. This concerns the exploration and extraction of ore and non-metallic materials, energy raw materials, water, products for mining chemistry, etc. The use of overburden at mining companies in the Kursk Magnetic Anomaly may be one of the examples of rational resource conservation[3, 4].

A reduction in the use of primary energy, the specific per capita consumption of which in the world in 2015 amounted to 1.85-1.90 tons of oil equivalentcan be considered the most important direction of resource saving [5]. Decrease of this indicator most actively affects many aspects of resource saving: volumes of extraction of raw materials and materials, transport costs, environmental costs, construction of various facilities, and much more. The main ways of solving them are known and are developing [5, 6]. However, for a number of countries, incl. Russia and Belarus, they are paramount for sustainable development.

The stage of primary preparation and processing of raw materials forms the entire subsequent cycle of technological production of materials. This is a multifaceted technology, but designintegrator redistributionformstheir backbone, which consumes about 10% of the total generated electricity and to 2% of the produced metalis lost for wearing [7].

At the same time, resource losses are no less great. The Granit mine of the city of Mikashevichi in the Brest region can serve as an example, where 18 million tons are extracted annually as a rock mass for granite grains, of which 4 million tons is a sifting of at least 5 mm in size and 95 % going into irretrievable losses. Such kind of resource waste is found everywhere and their practical realization has great prospects, but it is connected with many difficulties, mainly it concerns the organizational plan.

Increasing the degree of processing and extraction of minerals is a fundamental problem in the technological sphere of production and largely determines its effectiveness. This problem is exacerbated by the depletion of natural resources and the need to use finely divided and difficult-to-access minerals. One of the complex solutions in this direction can serve as selective ore preparation [7].

The problem of increasing the share of raw materials, materials and waste involved in production has a special attention. At the same time, we should talk about expanding the raw material base of the industry by developing new technologies for processing waste, utilizing harmful dust emissions, efficient water and other wastewater treatment, reasonable management of the forest fund, etc. In recent decades, the share of processed heterogeneous and complex in composition and properties of materials has increased significantly, which requires the creation of new technologies and equipment [8, 4].

Synthesis of new compounds and compositions is included in the sphere of advanced technologies and is based on modern advances in science and technology. Its range starts from nanotechnology and ends with the most multi-tonnage production, where more energy and materials are spent. The applied technologies are characterized here as obvious archaic and undoubted perfection, and in large-scale production they are characterized by stable conservatism and low intensity of development. The production of concrete products can serve as an example of the latest technology particularly in the processes of mechanical activation of concrete mixture components [9],

Resource-saving based on the increase the periods of operation of manufactured products should be regarded as the simplestway to save material resources of all kinds. Initially, it requires an increase in the cost of creating high-quality materials and products, but reduces initial production costs and subsequent operational costs. The reliability factor works successfully on non-durable products and is not determinative in the production of products for a market with high competitive offers or rush demand. Hence the conclusion-evaluation of this direction requires a differentiated approach and concrete justifications.

The sphere of related processes and factors includes:

- functional design;
- Optimization of building structures;
- technological energy saving and electric drive;
- technological transport of materials;
- use of renewable energy sources;
- Innovative management of the technological environment;
- personnel training.

The sphere of related processes and factors that also exert a certain influence on resource saving consists of directions that are described in detail in [6] and cover functional design of all types of technologies, optimization of building structures, technological energy saving, technological electric drive, technological transport of materials, use of renewable energy sources, innovative management of the technological environment, as well as training personnel training.

This is a purely subjective assessment of the non-productive sphere of technological resource saving, which allows us to expand our understanding of the potential and possibilities of this direction in addition to the production sphere of technological resource saving. Full-scale analysis of the technological environment requires additional research and collection of statistical, expert and other information.

In our opinion, it is necessary to highlight the most important problems and unsolved problems in the sphere of modern resource-saving:

- Increase the degree of processing and extraction of useful components;
- increase of the share of raw materials, materials and waste involved in the practical use;
- Synthesis of new compounds and compositions that ensure the usable use of raw materials, techno genic materials and waste;
- Optimization of technological processes that contribute to improving quality and saving expensive components;
- the discrepancy between the level of development of the technological machine-achievements in the field of materials science;
- the disparity of the scientific and technical and project base and corporate barriers in the sphere of technological resource saving;
- the absence of a single global center for the organization and management of the development of technological resource-saving.

At present, material science has moved far ahead, which allows creating new efficient materials, products and technologies, but their practical use is limited to extremely old technological equipment, control and management systems. World science has produced a large number of very different of anomalous effects in the processes of technological processing of materialsin nature and conditions of the occurrence.

However, in industrial technologies they are used extremely little. Thus, in particular, the mechanism proposed by the authors for the absorption enhancement of the strength of disperse structures, shown by the example of the introduction of solid natural adsorbents into concrete mixtures [10, 11],is not still used. It is assumed that after full-scale studies, its functions can go far beyond the proposed effect of Rebinder.

The most difficult situation with innovative development in the sphere of so-called production technologies is typical for Russia and Belarus. Here there is a paradoxical situation when foreign producers have wide access to our markets with very mediocre equipment and technologies, and their developers, capable and willing to create high-class equipment of the world level, do not have the opportunity to self-actualize in the conditions of the domestic market [12,13].

Conclusions. From the analysis of the above research and production tasks in the field of resource and energy saving technologies, it follows that the use of new forms of organization of creative teams that unite specialists of various profiles in their ranks can provide effective energy saving based on the creation of new materials, technologies, machines and complexes.

**Acknowledgments**

The article was prepared within the framework State task No. 9/11523/2018/11.12

**References**

[1] Demyanova VS, Kalashnikov VI, Kazina GN 2006 Ecological technical and economic aspects of the use of nonmetallic industrial wastes in the production of cement *Building Materials*vol 11pp 52-53

[2] SivachenkoL A, KusebayevU K ,. ReutskyI A, Rovsky A M 2015 Technological redistribution with maximum energy saving potential (Energy efficiency, Minsk) vol 10 pp 24-30

[3] Gibelev EI, Ilina TN 2009 Overburden rocks of mining enterprises in the technology of granulation of organic mineral fertilizers *Proceedings of the Tenth International Symposium of FSUE VIOGEM*(Belgorod: Publishing house VIOGEM) pp 337-342

[4] Ilyina TN 2009 Processes of agglomeration in the technology of processing of disperse materials: monograph (Belgorod: BSTU Publishing House) p 229

[5] Makarov A, GalkinaA, Grushevenko E 2014 Perspectives of the world ecology until 2040World Economy and International Relations vol pp 3-20

[6] Sivachenko LA, Sivachenko TL Technological engineering - an innovative reserve of the world economy: a monograph (Mogilev: Bel.-Rus. Univ.) p 254

[7] RevnivtsevVI, GaponovGV,Zarogatsky L P 1988 Selective destruction of minerals (M .: Nedra) p 286

[8] SevostyanovV S, ShinkarevL I, Sevostyanov MV and others 2011 Technical fundamentals of processing and utilization of man-made materials(Belgorod: BSTU Publishing House) p 263

[9] SivachenkoL A, SevostyanovV S, SivachenkoTL, Bagitova S Zh2016 Mechanoactivators of adaptive action and their development *Sb. sci. Art. Vibration technologies, mechatronics and controlled machines* vol 1(Kursk: YuSGU) pp 104-117
[10] Sevostyanov VS, Sivachenko L A, Ilina T N 2014 The main statements of physicochemical mechanics in the improvement of technological processes Ecology and rational nature management as a factor of sustainable development(Sat. tr. Belgorod: Publishing House of BSTU) pp 263-270
[11] Ilyina TN 2010 On the interaction of water-soluble components withcomponents of granulated charge materials *Vestnik of the Belgorod State Technological University named after. V.G. Shukhov* vol 2pp110-113
[12] Ilyina TN, Sevostyanov MV, Shkarpetkin EA 2010 Structural and technological perfection of aggregates for granulation of powder materials *Vestnik of the Belgorod State Technological University named after V.G. Shukhov*vol2 pp 100-102
[13] SevostyanovV S, Il'inaT N, SevostyanovM V, Shkarpetkin E A 2011 Investigation of the conditions of the microgranulation process in disperse systems*Vestnik of the Belgorod State Technological University named after. V.G. Shukhov* vol pp 81-86

# REDUCTION OF ENERGY CONSUMPTION DURING BUILDING MATERIALS PRODUCTION

**V V Gubareva**

Department of Heat Engineering Technology, Belgorod State Technological University of V.G. Shukhov, Belgorod, 308012, RU

**Abstract**. At present, in the production of building materials, which is one of the most energy-intensive industries, an increasingly urgent issue is the reduction in fuel consumption due to the use of secondary energy resources. The paper presents options for reducing energy costs through the use of heat from gases from furnaces in the ceramic industry.

The production of building materials occupies a leading position as an energy consumer and energy saving becomes the main task of the industry. On average, 1 ton of the finished product at building materials enterprises consumes 50 ... 100 kg. d.t.

The total heat utilization of the enterprise consists of useful heat perception of individual technological processes. The most important direction of energy saving is the creation of complex technologies ensuring the use of the potential of waste streams of various stages of the technological process [1].

Essential fuel economy, and as a consequence of reducing the pollution of the environment, can be achieved by utilizing the heat of waste gases, the use of which is still insufficient and an average of 30 ... 40%, while their share in general, the balance of possible use thermal secondary energy resources is 75% [2]. Furnaces are the largest industrial consumer of energy resources, and therefore energy-saving technologies must be used here [5].

Heat recovery systems for furnaces attract the greatest attention, especially waste heat, the so-called secondary energy resources. The possibility of using the heat of off-gases for technological and energy purposes depends on their temperature, the heat potential of the waste and the mode of their supply to the heat-using plant. In addition to the use of secondary energy resources, saving of consumption of primary energy resources is carried out by increasing the efficiency of process units and units [3].

In those industries where there are productive high-temperature installations, the step wise utilization of the temperature potential of the outgoing gases of furnaces and other aggregates allows in a number of cases to exclude the use of fuel for carrying out low-temperature processes, for example, drying [6].

In this paper, we discussed the possibilities of using the heat of gases emitted from furnaces in the ceramic industry.

The subject of the study was a tunnel kiln for burning clay bricks with a capacity of 17 million pieces per year. As a result of calculating the heat balance of the furnace, the temperature of the exhaust gases was determined to be 75 ° C and the flue gas volume was 10.68 $nm^3$ / $nm^3$. The clay bricks before the baking are dried in tunnel dryers using a low-potential drying agent.

The drying process is energy intensive and uneconomical in terms of heat[6]. An objective characteristic of the energy efficiency of drying plants is the specific consumption of thermal energy per 1 kg of evaporated moisture. Theoretically, the required amount of heat for evaporation of 1 kg of moisture is at the usual conditions of 2200 ... 2700 kJ / kg. The efficiency of drying plants is 30 ... 33%. An increase in the efficiency of drying installations would double the annual savings of up to 60 million tons of equivalent fuel [4].

At the same time, drying a brick in a tunnel dryer requires a drying agent with a temperature of the same order. But as a drying agent, gases that escape from the tunnel kiln are undesirable, since their use has a number of negative consequences:

- gas contamination of industrial premises is created because of looseness in dryers;
- metal part of trolleys is damaged because of metal corrosion;

- aluminum frames of trolleys are also damaged due to the presence of sulfur compounds in flue gases;
- quality of the dried bricks deteriorates.

Therefore, it is most advisable to use the heat of the waste gases to heat the drying agent [7]. The most effective way for this is often the use of regenerative heat exchangers. The industry uses regenerators with a rotating and fixed nozzle. Regenerators with fixed packing are regenerators of periodic action and are used at high temperatures of waste gases. Regenerators with a rotating nozzle are used mainly at low temperatures of exhaust gases and are devices of continuous operation.

Since for the dryer it is necessary to continuously supply the drying agent, we settled on a regenerative air heater with a rotating nozzle, which are used at low gas temperatures. The rotating nozzle consists of corrugated steel sheets. These regenerators are quite expensive, since their design consists of metal. Traditionally, vertical and horizontal regenerators are used in industry. Recently, sectional vertical regenerators are increasingly used. The main advantage of sectional regenerators in comparison with regenerators with a common chamber is that they can change the nozzle without disturbing the operation of the furnace.

The industry uses regenerators with rotating on-the-box three types: a nozzle in the form of a matrix, made of woven metal mesh (made of aluminum or stainless steel), a corrugated nozzle resembling honeycombs with a shallow cell and nonmetallic nozzles known as hygroscopic. The nozzle itself has a low cost, but at the same time it has a high thermal efficiency. The disadvantage of such attachments is a higher aerodynamic resistance. Also, unlike other types, they are more intensely contaminated.

To use the heat of the waste gases and to improve the efficiency of the heat technological scheme, a variant of a regenerative air heater with a second type nozzle was chosen, since it has less hydraulic resistance, allows reducing pollution and facilitate cleaning.

As a result of heat engineering and design calculations, we obtained that at a regenerator capacity of 40,200 $Nm^3$ / h, the necessary heat ex-change surface of 489,2 $m^2$, the nozzle volume of 0,245 $m^3$ and the mass of the nozzle is 1911 kg. The air temperature after the regenerator, with such parameters, reaches 57 ° C, which is quite acceptable for drying bricks. The fuel economy in the production of ceramic bricks was 23% with a payback period of less than a year.

The second option is the utilization of the heat of waste gases at a temperature of 200 ° C from the zone of heating of tunnel kilns in a spray dryer. This dryer is designed to produce a special morphological structure from the clay slip of a press powder, which is necessary for the semi-dry pressing of ceramic tiles. Drying in these devices is characterized by high intensity and the process ends extremely quickly (in about 15 ... 30 seconds). But in this case, despite the high temperature of the gases, the surface temperature of the material is only slightly higher than the adiabatic evaporation temperature, and the drying takes place under mild temperature conditions, obtaining a qualitative product that does not require further grinding [2].

As a result of the calculation of the spray dryer, the necessary volumetric flow of the coolant, the design dimensions of the dryer and the calculation and selection of auxiliary equipment were determined. According to the calculations made and the set values, a general purpose spray dryer with a diameter of 6,5 m was selected, a cylindrical section of 10 m in height, with a working volume of: $V = 330\ m^3$, equipped with 6 pneumatic nozzles. The results of the feasibility study fully confirm the feasibility of this option, because The unit cost of production decreased by 148 rubles / thousand units for brick, the payback period was 0,5 years.

**Acknowledgments**
The article was prepared within development program of the Flagship Regional University on the basis of Belgorod State Technological University named after V.G. Shukhov.

**References**

[1] Semenenko NA, Kuperman LI, Romanovsky SA 1979 Secondary energy resources and energy-technological combination in industry (Kiev: "Higher School") p 296

[2] Gubareva V V 2014 Some ecological aspects of utilization of secondary energy resources. Rational nature management as a factor of sustainable development (Belgorod: Publishing House of BSTU) pp 186 -192

[3] Danilov OL, Leonchik BI 1986 Energy saving during thermal drying(Moscow: Energoatomizdat) p 136

[4] Mikhailov VV, Gudkov Ya V, Tershchenko AV 1978 Rational use of fuel and energy in industry (M .: Energia)p 282

[5] Trubayev P A2007 Energy-technological analysis of high-temperature processes and apparatuses for the production of silicate materials*Bulletin of the Belgorod State Technological University named after V.G. Shukhov* vol 1 p 11-13
[6] Glagolev S N, Sevostyanov V S Gridchin A M2015 Thermal analysis of the technological line for extruding and drying technogenic fibrous materials *Bulletin of the Belgorod State Technological University named after V.G. Shukhov* vol 2 pp 78-83
[7] Trubaev P A 2006 Exergy analysis of thermal processes in the building materials industry Theoretical Foundations of Chemical Engineering vol 40(2) pp 175-182
[8] Trubaev P A, Tarasyuk P N 2017 Evaluation of energy-saving projects for generation of heat and heat supply by prime cost forecasting method *International Journal of Energy Economics and Policy* vol 7(5) pp 201-208

IOP Conf. Series: Journal of Physics: Conf. Series **1066** (2018) 012023 doi:10.1088/1742-6596/1066/1/012023

# RESEARCH AND DEVELOPMENT OF TECHNICAL MEANS FOR LIQUID OIL CLEANING

**V S Sevostyanov, L I Shinkarev and M M Belykh**

Department of Technological Complexes of Machines and Mechanisms, Belgorod State Technological University of V.G. Shukhov, Belgorod, 308012, RU

**Abstract**. This article deals with the problems of utilization of used oils, as well as ways of their regeneration. The methods as well as technical means for the regeneration of polluted oils were analyzed. Investigations were carried out on the researching the filtering materials and technical means for their implementation. Based on the conducted investigations, the authors propose to use a vibro-centrifugal unit for the regeneration of polluted oils.

Today, due to the increase in the number of cars, the consumption of oils is increasing all over the world, and as a result, the amount of waste in the form of usedoil is increasing, which is practically not utilized today, but is used as fuel or merges into water bodies and on the soil.

In Russia, averages of about 2-4% of used oils are processed. This is largely less than in most European countries, in which an average of 32 to 40% of used oil is processed. Therefore, the processing of used oils is relevant for our country [1].

The aim of the work is to analyze the existing methods and technical means of cleaning polluted oils to create a filtration unit.

An analysis of domestic and foreign literature shows that at present such methods as chemical-technological are widely spread where the polluted oil is treated in sulfuric acid [2]; The mechanical methods can be used forseparating solid particles (oil is passed through the filtering surface), passing oil through force fields and others [3]; Also there are electrostatic methods, which mean that in the unit oil is passed through an electric field (the process of electrophoresis). For better filtering, corrugated paper can be used [4].

To clean polluted oils, such units as the electrostatic cleaner ome-01-03, installations for the regeneration of foreign companies "PALL", and "Wotec", as well as filter presses of the "Saratov Plant of Technological Engineering" [5] are used.

In order to develop a technological scheme for the cleaning used oils, the main schemes for the processing and utilization of oils were considered.

Based on the investigations carried out, a plant was built for the regeneration of used oils. It is shown in Fig. 1.

Experimental studies of oil regeneration processes with the addition of adsorbent, sedimentation and subsequent filtration were carried out at this plant.

Before starting experiments with the adsorbent, experiments were carried out to find a filtering material for oil purification.

Semi-synthetic motor oil, SL / CF SAE 5W-40 [6], was used in the experiments.

In the course of experimental investigations, activated carbon, perlite, vermiculite, and cellulose were used as adsorbents. Various types of filters were tested, namely from perlite, vermiculite, activated carbon, paper, sawdust, wood flour and dust.

The search for a filtration element for the regeneration chamber was carried out. During the experiments, filtering elements such as perlite, vermiculite, activated carbon, crushed paper, wood flour were used [7].

At the first stage of the investigation, a method of static filtration (sedimentation) of polluted oils through various filter materials was used. The average filtration time of 200 ml of oil varied within 1.5-2 hours, and in the filtered oil an average of about 10-30% of the pollution was observed.

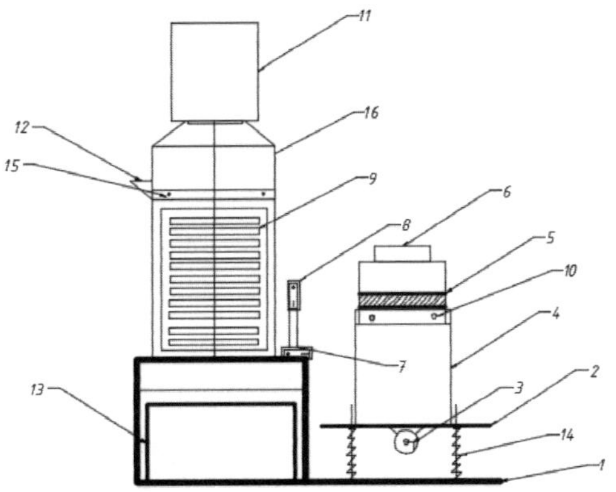


**Figure 1**. Scheme of the plant for the regeneration of used oils:
1 - bed; 2 - a vibrating platform; 3 - electric motor; 4 - the filter; 5 - filter elements; 6 - a jellied mouth; 7 - switch of the vibrating platform; 8 - the switch of the mixing plant; 9 - the blade; 10, 15 - flange connections; 11 - the electric engine; 12 - a receiving aperture; 13 –power supply block; 14 - springs; 16 - mixing capacity

In the future, in view of too long filtration time, vibration was used. At the same time, the number of turns of electric engine was about 2500-3000 rpm, and the amplitude of the oscillations was within (0.5-1) mm. A series of experiments showed that the filtration time decreased, on average by 30-50%, and was equal to about 50-90 minutes. The most effective filter elements were perlite, activated carbon, and pulp and paper wastes crushed in a hammer grinder and an increase of the pollutants number was observed in the filtered oil.

As the amount of pollution increased, we decided to use a double filter.

Filtering elements such as perlite and ground PPW (pulp and paper waste), vermiculite and ground PPW, activated carbon and ground PPW were used in the experiments. The average filtration time of 200 ml of working of averaged 120-140 minutes, but the amount of mechanical impurities and water decreased. As the filtration time increased, we decided to use the vacuum method. Three chambers were constructed, which are shown in Fig. 2. The average filtration time for each chamber (200 ml of oil) was about 10 minutes. In the first chamber, the greatest amount of mechanical impurities was observed but this number decreased. In the last chamber mechanical pollutions were practically not observed.

Chamber 1 Chamber 2 Chamber 3

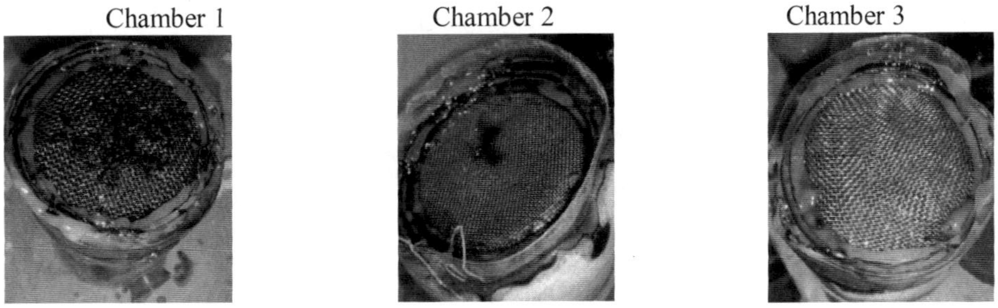

**Figure 2**. Filtration chambers

As a result of the obtained data, a 3-D model of the filter chamber was created for the vibration mixing unit whichhas the cranking speed was 250-400 rpm. The value of the crank eccentricity was 20 mm. The 3-D model of the chamber is shown in Fig. 3. Further, we created a special chamber for the model of this unit, we performed experiments on filtration with the use of vibration and vacuum methods. This model of the unit is shown in Fig. 4. After carrying out some experiments on a miniature plant, we obtained oil filtered from the mechanical impurities and water. The filtration time of 400 ml of the working of was an average of about 8 minutes, and according to the calculations, the average productivity of the unit for cleaning polluted motor oils was about 2.5 liters per hour.

The plant works as follows, the waste oil is poured into the filtration chamber (2), then the power is connected to the plant (5) and the vacuum cleaner (6)from the power supply unit(4), which functions as a pump, drawing air from the chamber through the hose (1 ).

Technical characteristics of the model:
shaft frequency ............................ 10-300 rpm;
amplitude of oscillation ............... 20 mm;
pump power ................................ 60 l / min;
created pressure ........................... ~ 2-4 atm.

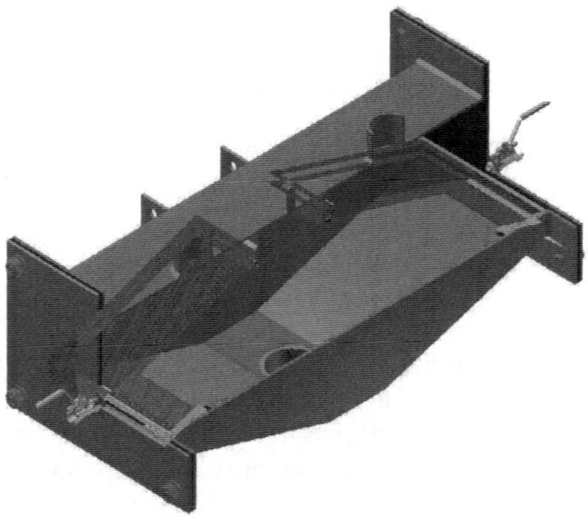

**Figure 3.** 3-D model of the chamber for cleaning polluted oils

**Figure 4**. Model of the unit for cleaning polluted oils:
1 - a branch pipe of the pump; 2 - filtration chamber; 3 - container with filtered oil; 4 - control unit; 5 - installation of vibro-centrifugalcleaning; 6 - pump

Using this model, it is possible to design a filtration unit that will restore the original properties of the oils by mixing the working of with the adsorbent and, after several stages of filtration, the oil will restore its original properties, which will allow it to be reused. Such a process unit for the cleaning of used automobile oils is multifunctional and can be used not only for waste oil cleaning, but also in the future for other polluted liquid media of related industries [8]. The main purpose of this unit is the regeneration of polluted oils.

For Russia, the development of such technological equipment will allow reducing the use of valuable resources of the country because working of materials will be used, the price of which is several times lower than the new ones.

**Acknowledgments**
The article was prepared within development program of the Flagship Regional University on the basis of Belgorod State Technological University named after V.G. Shukhov.

**References**
[1] Cleaning oil from mechanical impurities and water URL: http://ig.ua/ru/ochistka_ot_mekhanicheskikh_primesei_i_vody (reference date: 16/09/2017).
[2] ChuprovaL V, MullinaE R, ErshovaO V, Rodionov N I 2013 Chemical-technological aspects of the process of recovery of oil waste*The young scientist* vol 6 pp 235-237
[3] Bry IV, Kleimenova KF, Yakovleva ZI 1972 Regeneration of transformer oils: textbook(L : Chemistry Publishing House)p 168
[4] Tanesescu F, Kramaryuk R 1980 Electrostatics in engineering: Trans. with Romanians (M : Energia) p 296
[5] FuchsI G, EvdokimovA Yu, Lashkhi V L, Saidahmedov Sh M 1993 Ecological problems of rational use of lubricants (M .: Ed. Oil and Gas) p 158
[6] LUKOIL LUX Synthetic 5W-40 URL: http://lukoil-masla.ru/en/Products/ProductCard?product=52
[7] Kozhevnikov VP, Tokach Yu E 2015 Modern solutions for the processing solid domestic waste in BSTU *Vestnik of BSTU V.G. Shukhov* 2015 vol 1 pp 172 – 174
[8] SverguzovaSV, SakalovaGV, MalyovanyM S 2014 Efficiency of wastewater treatment of galvanic production by adsorption method*Vestnik of BSTU V.G. Shukhov*vol4 pp153-156

IOP Conf. Series: Journal of Physics: Conf. Series **1066** (2018) 012024 doi:10.1088/1742-6596/1066/1/012024

# RESOURCES AND PROBLEMS OF THE MATHEMATICAL SIMULATING THERMO-TECHNOLOGICAL PROCESSES

**V A Kuznetsov and P A Trubaev**

Department of Heat Engineering Technology ,Belgorod State Technological University of V.G. Shukhov, Belgorod, 308012, RU

**Abstract**. The paper presents the combustion of natural gas in a short flame in entourage of the very hot air had produced extremely elevated temperature of the flame, reaching 2300...2500 °C at the front of burning. In the stagnant part of the gas space beyond the short flame, the temperature of gases had gone down, approaching gradually to temperature of the molten-glass surface. As it is well known, a relatively low temperature level of the stagnant gas space allows zones of glass cooling to be organized at the end of the molten-glass basin of the glass furnaces with returned gas flow.

The contemporary methods of mathematical simulation allow rendering numerically and graphically the heat-transfer processes in high-temperature thermo-technological plants, what creates new possibilities for their researches. Mathematical models of the processes, executed in fluid medium, are focused on the numerical solving the complicated enough systems of partial differential equations. First of all, such are the Navier-Stokes equations, which numerical solution defines interrelated components of the fluid velocity $u$, $v$, $w$, directed accordingly along axes $x$, $y$ and $z$.

All the three Navier-Stokes differential equations for a steady subsonic flow can be presented in tensorial notations:

$$\frac{\partial \rho u_j u_i}{\partial x_j} - \frac{\partial}{\partial x_j}\left(\mu \frac{\partial u_i}{\partial x_j}\right) = -\frac{\partial \tilde{p}}{\partial x_i} + \frac{\partial \rho}{\partial x_i} gh, \qquad (1)$$

where $\rho$ is the fluid density; $g$ is the gravitational acceleration; $\mu$ is the dynamic viscosity, $h$ is a vertical coordinate. The static pressure $p$ is replaced here with a pressure function $\tilde{p}$, which introduces the gravity into the equations implicitly and uniformly for all the three coordinate axes [1]:

$$\tilde{p} = p + \rho g h,$$

Subscripts $i$ and $j$ accept values 1, 2 or 3 according to numbering of the coordinate axes. Besides, the agreement is implemented here that a sum of three terms of an identical structure can be represented in such an equation as one item with any subscript repeated in it.

The pressure function depends immediately on the three velocity components. As a result, a system of the interrelated differential equations would be formed. In order to solve them, the effective numerical double-sweep method has to be supplemented with numerous iterations, in order to specify factors in discrete counterparts of the differential equations.

The numerical solving the equation system becomes more complicated when the turbulent motion of fluid should be simulated. The eddy counterpart $\nu_e$ of the kinematic viscosity, subject to local values of kinetic energy of turbulence $k$ and of its dissipation rate $\varepsilon$, is usually determinedwith the help of standard turbulence model [2]:

$$\nu_e = C_\mu k^2/\varepsilon, \qquad (2)$$

IOP Conf. Series: Journal of Physics: Conf. Series **1066** (2018) 012024 doi:10.1088/1742-6596/1066/1/012024

where $C_\mu$is a factor of proportionality .

Values of $k$ and ε are computed at grid nodes by means of the numerical solution of two transport equations:

$$\frac{\partial \rho u_j k}{\partial x_j} - \frac{\partial}{\partial x_j}\left[\rho(\nu+\nu_e)\frac{\partial k}{\partial x_j}\right] = \rho(G-\varepsilon),$$

$$\frac{\partial \rho u_j \varepsilon}{\partial x_j} - \frac{\partial}{\partial x_j}\left[\rho\left(\nu+\frac{\nu_e}{\sigma_\varepsilon}\right)\frac{\partial \varepsilon}{\partial x_j}\right] = \rho(c_{\varepsilon 1}G - c_{\varepsilon 2}\varepsilon)\frac{\varepsilon}{k},$$

where $G$ is a rate of generation of the turbulence energy, which depends on the velocity distribution; $\sigma_\varepsilon$, $c_{\varepsilon 1}$, $c_{\varepsilon 2}$ are empirical factors.

It should be intended that the computer representation of numbershas a small error of their rounding, which considerably increases amassing in the double-sweep method for atotality of grid nodes. Numerical solving applied to the interdependent differential equations of mathematical models results inmultiple increasing faults of the computed functions.

For example, inexact values of the velocity that are transferred in the differential equations of the turbulence modelincreasethe inexactness of the eddy viscosity. Since the eddy viscosities have to be summed up with the kinematic ones in the equations (1), the fault of thevelocity components that would be computed in next iterations at certainlyinaccuratevalues of the eddy viscosity would be augmented. As a result, the iteration convergence might be slowed down or even ceased.

Such an algorithmhas beenapplied to the numerical simulation ofhaydite calcining in a rotary kiln, whose inner diameter equals 2 m, and the length is 40 m. The manufacturing 7 t/h of haydite consumed 700 $m^3$/h of natural gas at 10% excess of air. A calcined bed was formed in the furnace bywetclay granules,whose humiditymounted to 11 %. Air, heated up to 320 °C, was suppliedforthe fuel combustion through the unoccupied cross-section of the furnace.

Diffusion burning of natural gas was simulated by transport differential equations for the fuel and air concentrations. An additional transport equation for water vapor allowed computing a temperature reduction, produced in consequence of moisture evaporation from theraw granules. When calculating temperature of the internal hot surface of the walls,heat fluxesfrom gaseswere determined with taking account of convective and radiative transfer. Also the simulation took into considerationthe direct radiation of therefractory on the calcined bed in spectral "windows" andsudden changes ofthe wall temperature at periodicalentering of rotating walls under the bed.

In order to solve numerically the system oftwelve differential equations with a personal computer at more than 200,000 grid nodes, it had required to accomplish above 12,000 iterations, whathad ensured high accuracy of obtained results.

An image, presented at Fig. 1, adumbrates the temperature field in the kiln gaseous space.An average temperature of granules, which had reached 1200 °C in the heated-up bed, might be evidence ofa good compliance of thecomputedthermal conditions with technological requirements forthe haydite calcining. Thus,adequate resultshave been obtained at this mathematical simulation, in spite of the fact that the errors of the numberrounding slowed down the iterationconvergence.

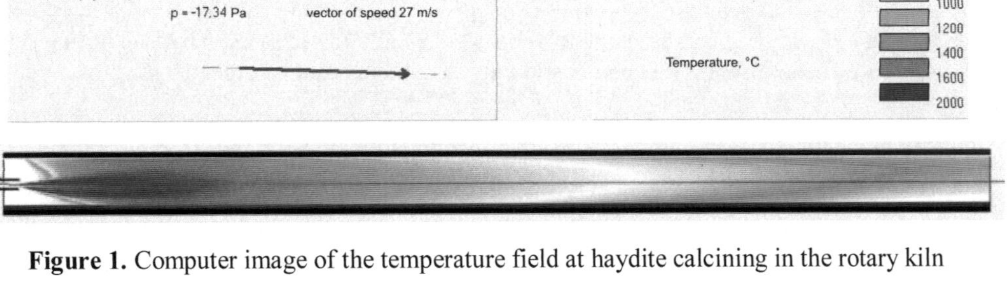


**Figure 1.** Computer image of the temperature field at haydite calcining in the rotary kiln

Nevertheless,some predicaments can appear sometimes inthe computer realizing mathematical models of high-temperature processes. There are anexperience of difficult numeric simulatingthe diffusion burning and heat exchange in a glass furnace, whose length was5,2 m and width 3 m. Fig. 2offera view onits front wall with two rectangular windows forinputting hot air and removal of gases. A longitudinal section of furnace, which is also presented there, givesanidea about a complicatedform of the computational domain.

The natural gas was supplied at rate of 180 $m^3$/h andvelocity of 102 m/s through a nozzle diameter of 25 mm in an inlet air duct, whose lower part constitutes a peculiar fire-chamber of 600 mmin length. The fuel nozzle is inclined up at angle 5°. The arch of the airductis established at angle 25° to the horizontal; because of that, a current of the air, which had been heated up in a regenerator to 1000 °C,got an impulse directed down to amolten-glass basin.

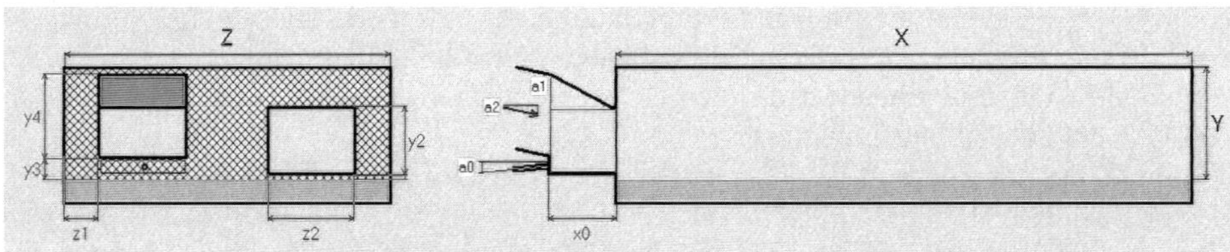


**Figure 2**. Outline of the glass furnace on the monitor screen for inputting initial data of the numeric experiment

A problem of choosing and applyingthe turbulence model that would be efficient enough becameespecially acute atthe numerical simulatingthe fuelburning in the furnace. The standard turbulence model did not assure the iterationconvergence insolvingthe system of interdependent equations for such an intricateflow of high-temperature gases. Calculation errors of the eddyviscosity produced by the formula (2) brought, finally, to unduly big fluctuations incomputational values of the velocity components,so thatthe iteration-convergencehad been ceased.

Therefore, calculation of theeddy viscosity had been performed by means of a new algorithm that used a modification of the formula (2):

$$\nu_e = C_\mu \sqrt{k\varepsilon/e}/e . \qquad (3)$$

Here $e$is a ratio of the turbulence-dissipation rate to the turbulence kinetic energy ($e \equiv \varepsilon/k$). Values of the function$e$have been defined by the solution of the differential equation [3] with supplementary items, which had not been accounted earlier.

Taking into consideration thatrelative errorsof valuesare summarized attheir multiplication and division,it is possible to demonstrate that calculation of the eddy viscosity by means of the formula (3) comes to be more precise.Since the ratio$e$had changed in iterations much weaker than quantities$\varepsilon$ and$k$, inaccuracy of the calculated eddy viscosity reduced almost three times as compared with that produced by the formula (2). Thishelped to restorethe iterationconvergence and allowed to realize numerically the mathematical model ofthe heat and mass transfer in the glass furnace with returned gas flow.

According to a curve in Fig. 3, the eddy viscosity of the gasincreasesby many timesat distance 1... 1,4 m from the air-inlet window in the front wall, owing toinhibition of the gas flow in a dead-end partof the furnace.Intensity of eddy mixing the fuelwith air was essentially raised there, what had been resulted inan extremely short diffusionflameof fuel combustion. Its computer image is presented inFig. 4.

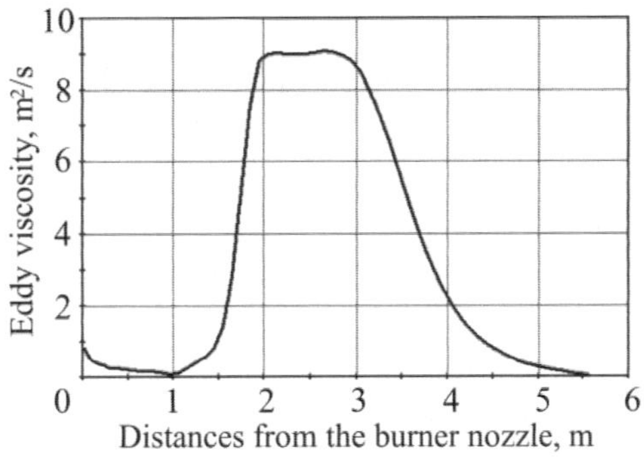


**Figure 3**. Change of the eddy viscosity of the gases along the furnace

**Figure 4**. View from above on the diffusion fame in the glass furnace with returned gases flow

Combustion of natural gas in a short flame in entourage ofthevery hot air had produced extremely elevated temperature of the flame, reaching 2300...2500 °C at thefront ofburning. In the stagnant part of the gas space beyondthe short flame, the temperature of gases had gone down,approachinggradually to temperature of themolten-glasssurface. As it is well known,a relatively low temperature level ofthe stagnant gas space allows zones of glass cooling to be organized at the end of the molten-glass basinofthe glass furnaceswith returned gas flow.

**Acknowledgments**
The article was prepared within development program of the Flagship Regional University on the basis of Belgorod State Technological University named after V.G. Shukhov.

*The article was prepared within the framework of the Base University Development Program on the basis of the BSTU named after V.G. Shukhov.

**References**

[1] Ferziger J H, Computational Methods for Fluid Dynamics 2002 *M. Perić. Berlin, etc: Springer*

[2] Patel V C, Rodi W, Scheuerer G 1985Turbulence Models for Near-Wall and Low Reynolds Number Flows *J. AIAA* vol 23 pp1308–1319

[3] Menter F R 1994Two-Equation Eddy-Viscosity Turbulence Models for Engineering Applications *J. AIAA* vol 32 No 8 pp1598–1605

# RESOURCE-SAVING TECHNOLOGICAL COMPLEX FOR POLYMER WASTE PROCESSING

**V S Sevostyanov[1], P Yu Goryagin[1],V VObolonsky[2]and R A Ermilov[1]**

[1]Department of Technological Complexes of Machines and Mechanisms, Belgorod State Technological University of V.G. Shukhov, Belgorod, 308012, RU
[3] Ltd TK Ekotrans, Belgorod, 308006, RU

**Abstract**. A technological complex for the production of composite mixtures from techno genic polymer materials is presented in the article. The development of resource-saving technology and special equipment is considered for complex processing of techno genic polymer materials with various physical and mechanical characteristics. The functions of the dispersed composition of the crushed polymeric materials under different conditions are constructed and rational operating modes of the rotor-centrifugal unit are established. To improve the efficiency and expand the technological capabilities of the rotor-centrifugal unit, a patent-protected construction of a combined-action separator with a step-by-step separation of material of various fractions has been developed.

The need to conduct research on the development of technological complexes for the processing of polymeric waste using the scientific and methodological foundations for the integrated utilization of techno genic materials is caused by the need to minimize the anthropogenic impact of industrial objects on the environment.

The aim of this work is to develop a technological complex for the processing of polymeric waste and to determine its optimum operating modes to minimize the anthropogenic impact of industrial objects on the environment.

In Belgorod region in Ltd "TK" Ekotrans" collection, sorting, processing and utilization of techno genic materials (TM) are organized. One of the technological tasks is to reduce the energy costs of the process and improve the quality of grinding of classified TM, in particular, PERT bottles [1, 2]. To solve this important research and production problem, we consider the developed technological complex for the production of composite mixtures from techno genic polymer materials (Fig. 1) [3].

The technological complex operates in the following way. The starting material is fed to the delivery tray 2 by means of a telpher 1. A conveyor belt 3 is mounted at the bottom of the tray. By means of a belt conveyor and a plate feeder 4, the material is fed to a rotor-centrifugal unit (RCU) 6. To trap metallic inclusions before the RCU an inclined tray with a magnet is established. In the upper part of the RCU, studded rolls are installed, to prepare PERT bottles for grinding, compressing and making air discharge holes in them. In the first coarse grinding chamber, the material is ground with disk cutters and, under the action of centrifugal forces, moves into the fine grinding chamber. There, the polymeric material is ground with wire brushes to a powdery state. After the RCU, the ground material is pumped to the separator by pneumatic conveying 7. The burden is divided into fractions by the action of air streams. Not fully ground material is deposed in the lower cone and returned to the regrinding in the second chamber of the RCU. The dusty fraction moves through the upper opening into the aspiration system.

The conditional material, deposited in the cyclones 8, enters the screw feeder 9. By means of the feeder 9 and the elevator 11, the raw material enters the bunker of the components of the composite mixtures 12. The bunker is equipped with partitions that separate it into three zones for separate, accumulation and storage of components. Before "Noria"a bunker of additional components of composite mixtures 10 is installed. From the bunker of additives 10, along the vertical elevator, the components are transported to the bunker compartments 12. Cell feeders are installed in the bottom of each compartment. From the bunker 12, the components enter the blade mixer 13, where they are mixed. After that, the composite mixtureis fed on the belt conveyor 14 for packing and weighing [4].

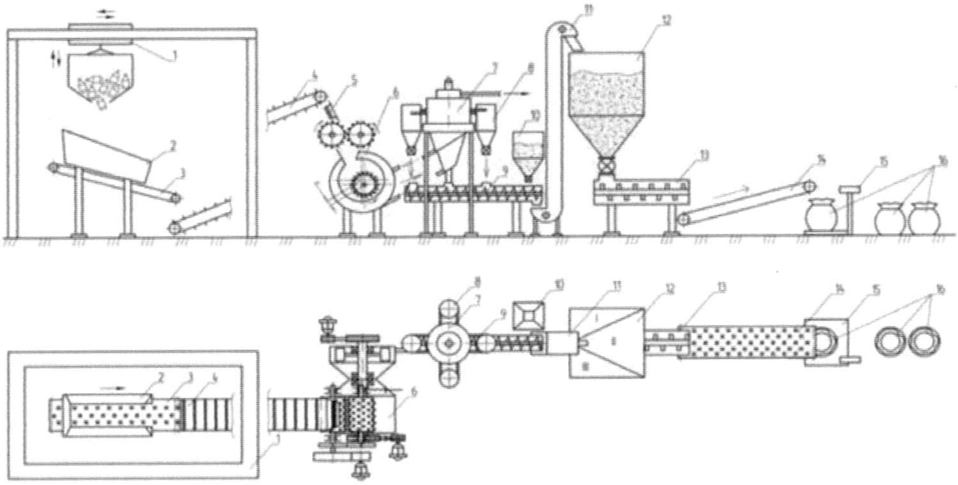

**Figure 1.** Technological complex of production of composite mixtures from techno genic polymer materials:
1 - telpher; 2 - delivery tray; 3, 14 - belt conveyor; 4 - plate feeder; 5 - the magnet; 6 - rotor-centrifugal unit; 7 - separator; 8 - cyclone; 9 - screw feeder; 10 - bunker of additional components; 11 - elevator; 12 - bunker of components of composite mixtures; 13 - the mixer; 15 - scales; 16 - big bag

The change in the granulometric composition of a material during its grinding was investigated in this paper. This is due to the need to use a separator that ensures the recirculation of material flows and their division into separate fractions.

To study the process of fine grinding of techno genic polymer materials in the second chamber of the RCU, some experiments were carried out. Hammer crusher in testing experiments is used as an analogue of the second chamber of the RCU with a wire brush rotor [5]. The crushed material was weighed and divided into portions of 100 gr. There werefollowing of experiments on grinding PET bottles with a hammer crusher. A mesh with an opening diameter of 2 mm was installed in the crusher. The material was crushed during 1, 4 and 7 minutes. To determine the fractional composition of the material, after each experiment, the crushed material that has fallen into the bag and the material that remained in the crusher chamber were collected separately, sieved on sieves with different opening diameters and weighed (Figure 2-4) . In the experiments, sieves with opening diameters of 2, 4, 7 and 10 mm were used. According to the results of the experiments, the functions of the dispersed composition of the crushed material for different grinding time were constructed (Fig. 5). It can be seen from the graphs that when the grinding time of the material increases, the size of the material fractions decreases. In the course of the experiments it was established that the crushed material, whose size exceeds the diameter of the grid openings, in view of its elastoplastic properties, passes through the mesh. In this case, part of the finely groundproduct in the crusher is trapped in the aspiration system.

**Figure 2.** Material with a particle size of 4 mm $\leq$ d <7 mm

**Figure3.** Material with a particle size of 2 mm ≤ d <4 mm

**Figure 4.** Material with particle size d <2 mm

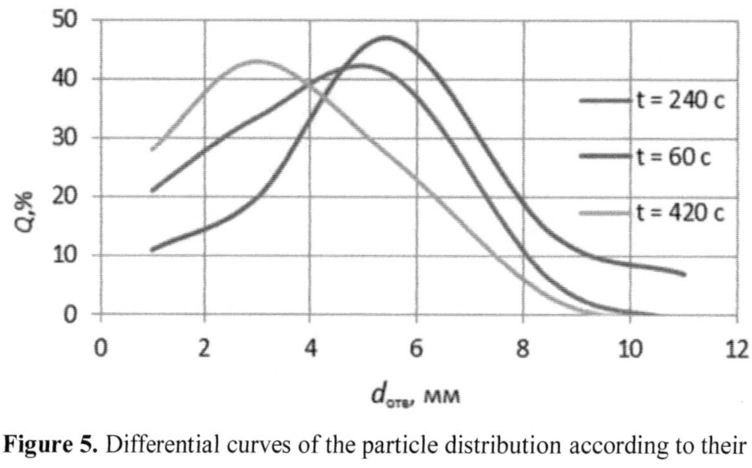


**Figure 5.** Differential curves of the particle distribution according to their size

To increase the efficiency of the grinding process and return to the re-grinding of particles passing through the hammer crusher grid, it is necessary to use a separator together with the RCU.

The design of the combined action separator used with stepwise separation of material from different fractions (fig 6) consists of two steps. The first stage serves for the isolation of a coarse product, the second for a fine product [6]. The separator works in the following way. The initial gas-material flow enters tangentially into the lower truncated cone 6 on the loading branch 8. Ascending along a spiral along the ascending line, the largest particles of the mixture are pressed under the action of a centrifugal force to the inner surface of the truncated cones 5 and 6. Particles lose their speed due to the increasing section of the classifier and fall into the grooved channel. The unloading of the product is carried out through the discharge pipe 7. The degree of classification of the mixture can be regulated by changing the pressure of the compressed air. To ensure the free flow of particles through the discharge pipe 7, the angle of inclination to the vertical axis must coincide with the angle of the natural slope of the material. Entering the upper part of the classifier housing, the fine particles along the tangentially integrated nozzles enter the remote elements 14, where the gas material stream rotates in the annular channel. The bulk of the particles trapped in the central chamber passes from it together with the gas material stream into the part of the upper chamber 1. From where the smallest particles, under the influence of excess pressure and additional rarefaction, created by the electric actuator of the separator 3 , enter the dust collector 9 and then to the cylindrical mesh frames 12 and the muffled sleeves 13. A baffle with holes 11 installed in the upper part of the dust collector 11, made of a gas-impermeable material, prevents the entry of a large number of material particles into the upper part of the classifier. This, in turn, prevents them from entering the atmosphere, and the openings made on it serve to pass the gas flow and create an additional rarefaction by the electric drive of the rotor 3. The particles caught in the dust collector 9 are discharged from it through the discharge pipe 10.

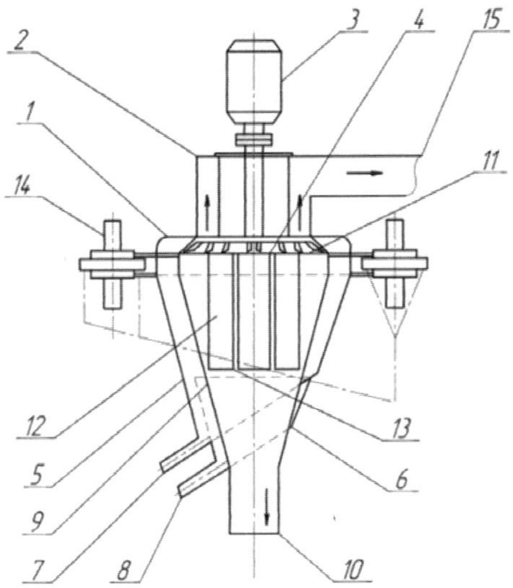


**Figure 6**. General view of the separator: 1 - upper part of the casing; 3 - electric drive of the separator; 4 - impeller with blades, 5, 6 - upper and lower truncated cones, respectively; 7, 8 - unloading and loading pipes, respectively; 9 - a dust collector; 10 - central discharge pipe; 11 - partition with holes; 12 - cylindrical mesh frames; 13 - fabric sleeves; 14 - remote elements; 15 - clean gas pipeline.

Thus, we have developed a resource-saving technology and special equipment for complex processing of techno genic polymer materials with various physical and mechanical characteristics. A design for combined grinding has been developed. According to the results of studies of the operating modes of the RCU, the functions of the dispersed composition of the ground polymeric materials under different conditions were built, which allowed establishing rational modes of operation of the RCU and its efficiency at different grinding time. To improve the effectiveness of the RCU, a patent-protected construction of a combined-action separator with a step-by-step separation of material from various fractions was developed.

The application of the RCU in combination with a closed-cycle separator allows expanding the technological possibilities:

- regulation of the aerodynamic parameters of the separator with the possibility of using polydisperse materials with different granulometry and density;
- the possibility of air cooling of the crushed polymeric materials and prevention of their spontaneous thermoplastic aggregation due to a decrease in their adhesion capacity.

**Acknowledgments**

The article was prepared within development program of the Flagship Regional University on the basis of Belgorod State Technological University named after V.G. Shukhov.

**References**

[1] Arashkevich D A 2003 Recycling of plastic waste and special rotary crushers *Plastic masses* vol 3 pp 5-13
[2] Ivanova O A , Rekhovskaya E O 2015 Recycling and processing of plastic bottles *Young Scientist* vol 21 pp 54-56
[3] SevostyanovV S, UralV I, SevostyanovM V, NosovOA 2015 Technological complexes and equipment for processing and recycling of technogenic materials: textbook (Belgorod: BSTU Publishing House) p 321
[4] Suprun LV, Romanenko S V, Tsygankova T S 2012 Analysis and solution of the problem of recycling and recycling of polyethylene terephthalate (PET) wastes *Vestnik of Siberian Science* vol 4pp 107-112
[5] Sevostyanov M V 2016 Resource-saving equipment for complex processing of technogenic materials *Bulletin of Belgorod State Technological University named after V G Shukhov* vol 4 pp 140-145
[6] Gridchin A M, Sevostyanov V S, LesovikV S, and others Pat. RU 2264268 C1 Ros. Federation, the IPC in 07 V 7/086 Vortex acoustic classificator; *The applicant and patent holder of BSTU V G Shukhova* No. 2004123996/03; claimed. 5.08.2004; publ. 20.11.2005, Bul.32p 11

IOP Conf. Series: Journal of Physics: Conf. Series **1066** (2018) 012026 doi:10.1088/1742-6596/1066/1/012026

# STUDY OF POWER DEPENDENCE OF WIND POWER FROM WIND SPEED

**K P Trubaev**

State Municipal Budget School No.45, Belgorod,308015, RU

**Abstract.** The paper presents aim of the work was to study the dependence of the power of the wind generator on wind speed on the laboratory installation. The hypothesis is proved - the higher the wind speed, the greater the power of the wind generator. But at low wind speeds, the wind generator rotates, but does not generate electricity. In Belgorod, the average wind speed is about 4 m / s, so wind generators will work, but not full power.

The needs of mankind for electricity are increasing every year. At present, most of the world's electricity is generated by power plants that burn irreplaceable fuel - natural gas, oil and coal. The fuel reserves are not unlimited, they can end in a few decades.

When burning fuel, harmful substances are released, which adversely affects the environment. With the goal of eliminating these problems, alternative sources of energy have been used in the world.

One of the promising areas of alternative energy is the use of wind turbines [1]. The wind generator works thanks to the wind, which rotates its blades. The energy of rotation of the shaft is transferred to the electric generator, as a result, electricity is generated.

In the course of studying and theoretical analysis of literature, we were interested in the following question: does the power generated by the wind generator depend on the wind speed?

Thus, the purpose of the work is to study the dependence of the power of the wind generator on wind speed.

Object of work: power of wind generator.

The subject of the study is the dependence of the power of the wind generator on the wind speed.

To achieve the goal, the following tasks were accomplished:

1) scientific sources of information have been studied and basic concepts have been revealed on the question of research [2-4];

2) the necessary equipment for carrying out a physical experiment has been prepared;

3) conducted laboratory studies aimed at studying the dependence of wind turbine-toroot wind speed;

4) the results of studies are processed and systematized:

5) investigated what the wind speed in Belgorod;

6) a comparative analysis of the results obtained

Empirical, practical and theoretical research methods were used in the work. The basic methods (study and analysis of literature and Internet resources, measurement, physical experiment, comparison of values, construction of tables and graphs) made it possible to determine the dependence of wind turbine power on wind speed and to determine that in Belgorod the wind generator will not work at full capacity.

The practical stage included the conduct of laboratory studies aimed at studying the dependence of wind turbine power on wind speed.

For the study, a special installation was used, consisting of: a fan, a wind generator, an anometer, a tachometer, a wattmeter (Fig. 1).

**Figure 1**. Laboratory installation

The fan, located opposite the wind generator, created a flow of air. The aneometer, installed in front of the wind generator, measured the velocity of the air flow. The tachometer, on a special stand, measured the rotor speed of the wind generator. A wattmeter connected to a wind generator measured the power of an electric current.

After the blades of the wind generator were untwisted, the instrument readings were taken. The experiment was carried out at different airflow rates. Based on the results of the experiment, graphs were constructed (Fig. 2, 3).

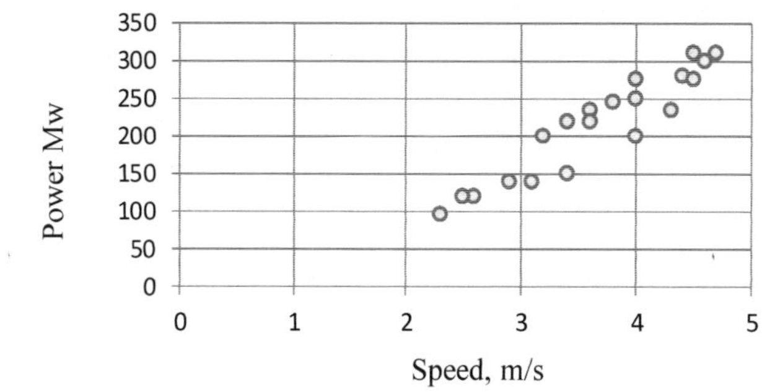


**Figure 2.** The dependence of the generated power on air speed

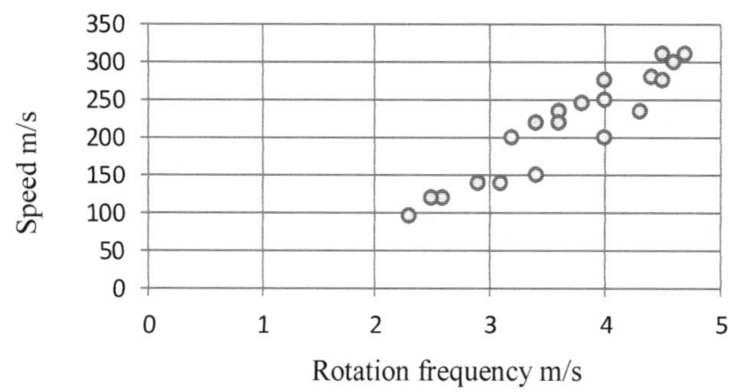


**Figure 3.** Dependence of the generated power on air speed

From the analysis of graphs, it can be concluded that at a flow rate of less than 2 m / s, despite the rotation of the wind generator, no electricity is generated.

In the further analysis of the work of wind generators in Belgorod. To do this, we used the data of the "Reliable Prognosis" site [5], which contains the data archive of the meteorological station of the city of Bel-city. Data on wind speed for 10 days of January and 10 days of July 2017 are shown in Fig. 4.

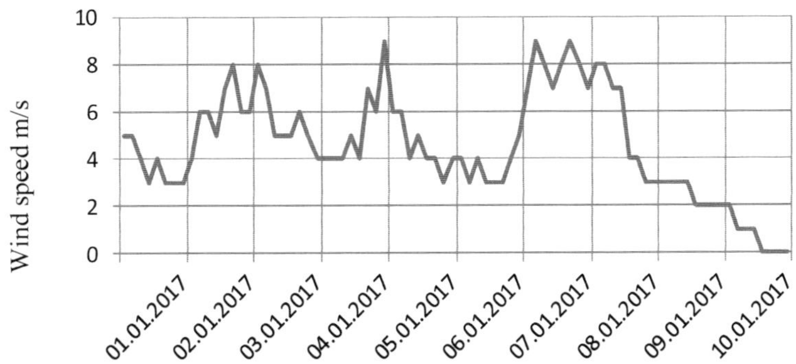


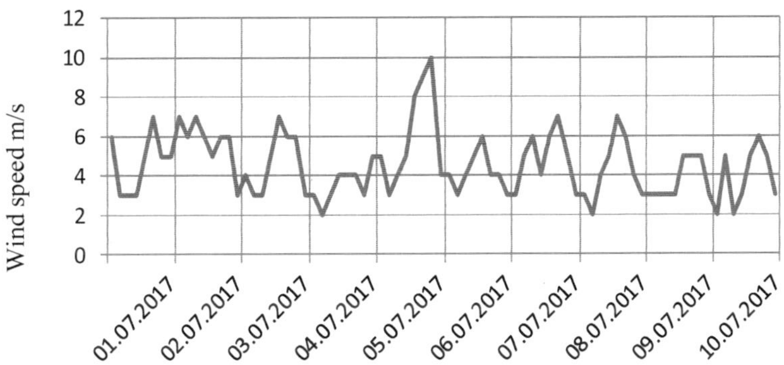


**Figure 4.** Change in wind speed during 10 days

By setting the start and end date on the site you can get average indications for the period. The average wind speed for seven years and average speed for each month for the last year were investigated (Tables 1, 2).

**Table 1.**Average and maximum wind speed for the year, m / s

| Year | Average | Maximum |
|---|---|---|
| 2011 | 4,2 | 15 |
| 2012 | 4,4 | 12 |
| 2013 | 4,4 | 10 |
| 2014 | 4,1 | 15 |
| 2015 | 4,3 | 14 |
| 2016 | 4,0 | 10 |
| 2017 | 4,2 | 14 |

**Table 2**.Average and maximum wind speed per month, m / s

| Month,year | Average | Maximum | Month,year | Average | Maximum |
|---|---|---|---|---|---|
| January 2017 | 4,3 | 10 | July 2017 | 3,9 | 16 |
| February 2017 | 4,7 | 10 | August 2017 | 3,8 | 10 |
| March 2017 | 4,4 | 14 | September 2017 | 4,1 | 13 |
| April 2017 | 4,0 | 13 | October 2016 | 3,8 | 10 |
| May 2017 | 4,1 | 9 | November 2016 | 4,8 | 10 |
| June 2017 | 4,6 | 10 | December 2016 | 4,3 | 9 |

It can be seen from the weather archive that the wind speed in Belgorod is constantly around 4 m / s. Therefore, wind generators will work, but not at full capacity.

Investigating this topic, you can make conclusions:

Firstly, after studying various sources of information, the basic concepts on research were revealed and it was determined that the higher the wind speed, the greater the power of the wind generator.

Secondly, it can be concluded that in the course of laboratory studies it is possible to study the dependence of the power of the wind generator on wind speed.

In the course of the work, the main goal was achieved-the dependence of the power of the wind generator on wind speed was studied.

Proofed the hypothesis - the more the wind speed, the greater the power of the wind generator. But at low wind speeds, the wind generator rotates, but does not generate electricity. In Belgorod, the average wind speed is about 4 m / s, so wind generators will work, but not full power.

**Acknowledgments**
Scientific work is the winner of the regional competition for students of educational institutions of the Belgorod region on electrical safety and energy efficiency "Energy and Man" (2017)

**References**:

[1] KhakhalevL V2008 Non-traditional and renewable sources of energy: *A manual for conducting practical work. lessons* (Ulyanovsk: USTU)p 32
[2] NatureTime - The site of ecological literacy (WebSite)
[3] Portal *Alternative Energy* (WebSite)
[4] Wind turbines: questions and answers (WebSite)
[5] The weather schedule (WebSite)

# TECHNICS AND TECHNOLOGY OF PROCESSING BASALT WASTE

**V S Sevostyanov , A V Shatalov, M V Sevostyanov , V S Krutov and V A Shatalov**

Department of Technological Complexes of Machines and Mechanisms, Belgorod State Technological University of V.G. Shukhov, Belgorod, 308012, RU

**Abstract**. During the processes of the production of mineral cotton, a large proportion of non-standard waste products are produced. Besides the costs on their processing, the costs on their processing and disposal (transportation, placement, conservation, etc.) are also required. In addition, waste has a harmful effect on the environment. All this predetermines the need of carrying out the research aimed at studying the physical and mechanical characteristics of materials, the possibilities and methods for their secondary processing, and use in the production of heat and resource-saving products.

One of the important areas of resource and energy saving is the provision of heat insulation of industrial and residential buildings due to a constant increase of costs on their heating. As a result of this, the demand for environmentally pure insulants significantly increases, under the production of which various heat insulating materials are used, including - basalt fibers [1].

In existing methods of producing basalt fiber, waste can be up to 15% during production, which is most often simply disposed of. All this undoubtedly has a detrimental effect on the environment.

The purpose of our research was to find out possible ways of using waste in the production of basalt fiber, to study its physical and mechanical characteristics, and to determine the optimum fractional composition for use in finished products [2]. There was also the task of creating an apparatus and a method for deagglomerating technogenic fibrous materials [3].

Basalt fiber is obtained from the basalt melt in smelters by free flowing through spinnerets made of heat-resistant metals. Since raw materials for the production of rock cotton are rocks, due to this heat-insulating products meet the requirements of fire safety and can be used as fire insulation and fire protection. The limit of the agglomeration temperature of basalt fibers is 1000C, which causes high values of the maximum operating temperatures [4].

In the process of manufacturing mineral cotton, in addition to cutting areas of the edge of the carpet (6 ... 8%), additional non-standard and unsalable goods (5 ... 8% of the total output) are formed. Thus, from 5 to 15% of products of total production due to their condition are not suitable for implementation and use. In many cases, it is impossible to recycle the forming amounts of trimming edges, and the residue has to be disposed of as waste.

Defective products and trimming edges are exported for disposal to special landfills. It has a detrimental effect on the environment in its turn. In addition, the material particles obtained during the processing of basalt fiber can enter the human lungs and cause undesirable consequences. For this purpose, the binding components are introduced into the composition, which makes it possible to create a relatively safe main product, for using it under preparing heat-shielding materials. Nevertheless, artificial building materials may contain residual components, which adversely affect the human body.

Basalt fiber, in addition to the above said, is an effective micro-reinforcing additive in concretes and other solutions on a cement or gypsum basis. It increases the resistance of concrete to deformation at the initial stage. When concrete begins to shrink (at a later stage), basalt fibers prevent cracking of concrete, thereby reducing the risk of breaking. In production of decorative products from cement or gypsum the basalt fiber reduces the number of defective of products by 90 ... 100%.

The use of basalt fiber in concrete solutions reduces the formation of shrinkage cracks at an early stage by 90% (the reinforcing grid reduces this figure by only 6%). But, in monolithic housing construction, basalt fiber cannot yet be used as a substitute for steel reinforcement. In this way it is permissible to use reinforcement from basalt plastic.

The use of basalt fiber allows slowing down the process of hydration of concrete, thereby reducing internal stresses at temperature fluctuations. Basalt fiber is resistant to physical damage during mixing, as well as to all chemical substances that make up concrete and alkalis used in production processes. It is also thermo stable and is not subjected to corrosion (which is typical for steel fiber), easily distributed without forming clots, durable, compatible with any additives and admixtures in concrete, plasticizers, antifreeze additives, hardening accelerators and retarders of solidification[5].

Based on the above described analysis of the areas of use of basalt fibers, there is not only the task of providing the necessary technological conditions for the processing of techno genic materials, but also the obtaining micro fibrillated fillers from them with given physical and mechanical characteristics: size, shape, dispersion, etc.

To solve the problem of recycling basalt fiber products waste it is proposed to use it for dispersion reinforcement concrete. We conducted patent studies to find the technical solutions used to disaggregate waste from the production of basalt fiber [6].

The technical proposal is shown in Fig. 1.

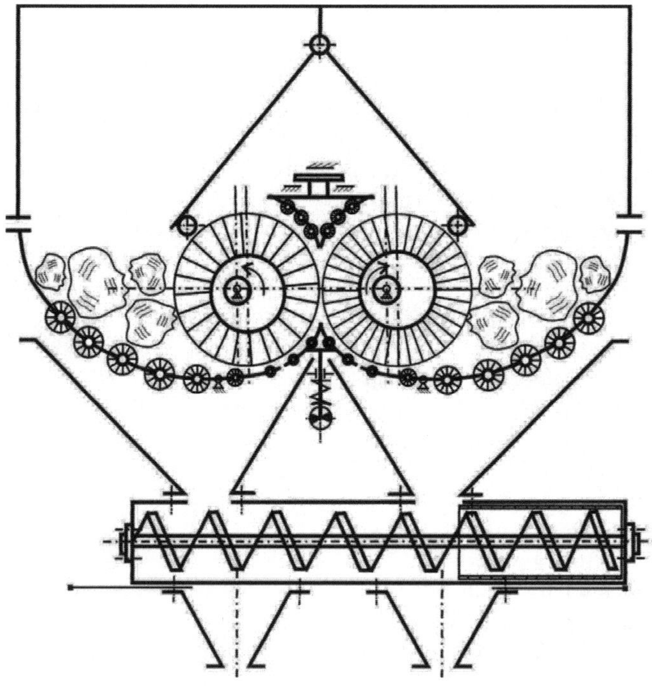

**Figure1**. Device for disaggregate

The method of disaggregate of techno genic fibrous materials consists in repeated impact of wire brush working elements fixed on rollers. Rollers having different circumferential speeds provide tensile forces and act on the pressed fibrous materials. The device allows the screening of fine-grained materials, and also due to vibration, eliminates the compaction of materials between rotating wire brush rollers towards each other and facilitates the spatial orientation of the fluffy fibers in the direction of flow. The method also provides for the selection of the disaggregated material according to separate fractions.

Thus, the method and construction of the device we propose will separate effectively the fibrous materials and also make the choice of disaggregated material according to the necessary separate fractions. The resulting basaltic fiber can be used as an effective micro-reinforcing additive in concretes and other solutions. All this will solve the problem of utilization of basalt waste and reduce production costs.

**Acknowledgments**
The article was prepared within development program of the Flagship Regional University on the basis of Belgorod State Technological University named after V.G. Shukhov.

**References**

[1] Uralsky V I, Sinitsa E V, Uralsky A V, Sazhneva E A 2017 Technological module of the closed grinding cycle *Vestnik of BSTU V.G. Shukhov* **10** pp 144-148

[2] Sevostyanov M V 2016 Resource-saving equipment for complex processing of technogenic materials *Vestnik BSTU named after V.G. Shukhov* vol 4 pp140-145

[3] Sevostyanov V S 2006 Energy resource-saving grinding units Monography (Belgorod: BSTU Publishing House) 2006 p 436

[4] Basalt technologies URL: http: // www. basaltech.org/ forum / forum73 / topic481 / messages

[5] Construction materials URL: http://budmat.kiev.ua/stattjy/statjy/bazaltovayafibra.html.

[6] Shatalov A V, Dubinin N N, Shevlyakov V I 2014 Energy-saving grinding complex Ecology and rational nature management as a factor of sustainable development: Sat. doc. Intern. scientific-techn. (Conf. Belgorod: BSTU Publishing House) pp 363-366

IOP Conf. Series: Journal of Physics: Conf. Series **1066** (2018) 012028 doi:10.1088/1742-6596/1066/1/012028

# TO THE QUESTION OF BIOGAS COMBUSTION IN VORTEX BURNERS

**D Yu Suslov, R S Ramazanov and M A Shvydkaya**

Department of Heat and Gas Supply and Ventilation, Belgorod State Technological University of V.G. Shukhov, Belgorod, 308012, RU

**Abstract**. The introduction of bio gas in the heat-generating power industry is impossible without the development and modernization of the main and auxiliary boiler equipment, as well as burner devices. The possibility of using the GGV-100 vortex gas burner for burning bio gas with a reduced methane content is considered. The main parameters of gas fuel are determined: the required volumes of oxygen and air for complete combustion of bio gas, theoretical and actual volumes of combustion products, the value of the maximum normal flame propagation velocity. In the software package Solid Works Flow Simulation, a computer model of the GVV-100 vortex gas burner was developed, as a result of the calculation, patterns were constructed of the change in the speed of the gas-air mixture in the longitudinally vertical plane of the section of the gas burner-furnace

Natural gas is the most efficient and environmentally friendly source of energy, which is the main fuel for energy supply of a huge number of consumers of various types, including agricultural enterprises. One of the most promising and effective areas in the energy supply of agricultural enterprises is the production and use of a renewable energy source - bio gas. Biogas consists of 30-80% methane, 20-50% carbon dioxide, and a small amount of other gases (hydrogen sulfide, nitrogen, oxygen, hydrogen, ammonia) [1]. Since the methane content in bio gas starts with only 40% of the total gas mixture volume, and the gas burners of most devices are designed for burning natural gas with a methane content of more than 90%, there is a need for research on the development of gas burners for burning bio gas fuel.

The purpose of the work is to determine the main energy parameters of bio gas fuel and study the process of formation of a gas-air mixture in low-pressure vortex burners.

When using bio gas of various compositions in gas burners, it is necessary to supply an amount of air that ensures optimal combustion of the gas-air mixture formed as a result of mixing of the gas used with the supplied air.

Let's calculate the theoretical volume of dry air for combustion of bio gas with the following conditional composition: $CH_4$ = 45%, $CO_2$ = 52%, $N_2$ = 1%, $H_2S$ = 2%.

The consumption of oxygen and air is determined on the basis of the combustion equations of the components that make up the bio gas [2].

Methane $CH_4 + 2O_2 + 7.52\ N_2 = CO_2 + 2H_2O + 7.52\ N_2$;

Hydrogen sulfide $H_2S + 1.5O_2 + 5.64N_2 = SO_2 + H_2O + 5.64N_2$.

On the basis of the above combustion equations for components of gaseous fuel, the volume of oxygen required for combustion of 1 m3 is determined by the formula:

$$V_{O_2}^{o} = 0{,}01 \cdot \left[ 0{,}5CO + 0{,}5H_2 + 2CH_4 + 1{,}5H_2S + \Sigma\left(m + \frac{n}{4}\right)C_mH_n - O_2 \right] \tag{1}$$

$$V_{O_2}^{o} = 0{,}01 \cdot [2 \cdot 45 + 1{,}5 \cdot 2] = 0{,}93m^3 \tag{2}$$

The volume of air $V_{air}{}^{\circ}$, which is necessary for the combustion of gas, is determined from the air content of 21% of oxygen by volume, $m^3$ / $m^3$ of gas:

$$V_в^o = \frac{100}{21} \cdot V_{o_2}^o = 4{,}76 \cdot V_{o_2}^o, m^3 \quad (3)$$

$$V_в^o = 4{,}76 \cdot 0{,}93 = 4{,}43 m^3 \quad (4)$$

The theoretical volume of dry trihydric gases, m3/m3:

$$V_{CO_2}^0 = 0{,}01 \cdot (CO_2 + CO + \Sigma m C_m H_n) \quad (5)$$

$$V_{SO_2} = 0{,}01 \cdot H_2S \quad (6)$$

$$V_{CO_2}^0 = 0{,}01 \cdot (52 + 45) = 0{,}97 m^3 / m^3 \quad (7)$$

$$V_{SO_2} = 0{,}01 \cdot 2 = 0{,}02 m^3 / m^3 \quad (8)$$

Theoretical yield of nitrogen in combustion products:

$$V_{N_2}^0 = 0{,}79 \cdot V_в^0 + 0{,}01 \cdot N_2 \quad (9)$$

$$V_{N_2}^0 = 0{,}79 \cdot 4{,}43 + 0{,}01 \cdot 1 = 3{,}51 m^3 / m^3 \quad (10)$$

Theoretical yield of water vapor in the combustion of gaseous fuels:

$$V_{H_2O}^0 = 0{,}01 \cdot \left( H_2 + H_2S + \Sigma \frac{n}{2} C_m H_n + 0{,}124 \cdot d_г \right) + 0{,}0161 \cdot V_в^o \quad (11)$$

where d is the absolute humidity of the gaseous fuel, g / m$^3$ (we take 5 g / m$^3$ [3].

$$V_{H_2O}^0 = 0{,}01 \cdot (2 + 2 \cdot 45 + 0{,}124 \cdot 5) + 0{,}0161 \cdot 4{,}43 = 0{,}99 \text{ m}^3/\text{m}^3 \quad (12)$$

The theoretical volume of dry gases consists of three-atom gases and nitrogen:

$$V_{с.г.}^0 = V_{RO_2}^0 + V_{N_2}^0 \quad (13)$$

where

$$V_{RO_2}^0 = V_{CO_2}^0 + V_{SO_2}^0 = 0{,}97 + 0{,}02 = 0{,}99 m^3 / m^3 \quad (14)$$

$$V_{с.г.}^0 = 0{,}99 + 3{,}51 = 4{,}5 m^3 / m^3 \quad (15)$$

The total theoretical volume of combustion products is:

$$V_г^0 = V_{с.г.}^0 + V_{H_2O}^0 \quad (16)$$

$$V_г^0 = 4{,}5 + 0{,}99 = 5{,}49 m^3 / m^3 \quad (17)$$

The actual volumes of combustion products are calculated with considering the indicator of excess air $\alpha = 1{,}2$.

Due to the fact that the process of oxidation of combustible fuel compounds is basically completed in the combustion chamber, the volume of tri-gas gases $V_{RO_2}$ remains unchanged throughout the flue gas path. The increase in the volume of dry gases is due to the volumes of nitrogen and oxygen of excess air :

$$V_{N_2} = V_{N_2}^0 + 0{,}79(\alpha - 1) \cdot V_в^0 \quad (18)$$

$$V_{N_2} = 3{,}51 + 0{,}79(1{,}2 - 1)4{,}43 = 4{,}21 m^3 / m^3 \quad (19)$$

$$V_{O_2} = 0{,}21(\alpha - 1)V_в^0 \quad (20)$$

$$V_{O_2} = 0{,}21(1{,}2 - 1)4{,}43 = 0{,}19 m^3 / m^3 \quad (21)$$

Consequently, the volume of dry gases can be recorded:

$$V_{d.g.} = V_{RO_2} + V_{N_2} + V_{O_2} \quad (22)$$

$$V_{d.g.} = 0,99 + 4,21 + 0,19 = 5,39\, m^3 / m^3 \quad (23)$$

Due to excess air, the volume of water vapor also increases somewhat:

$$V_{H_2O} = V^0_{H_2O} + 0,0161(\alpha - 1)V^0_в \quad (24)$$

$$V_{H_2O} = 0,99 + 0,0161(1,2 - 1)4,43 = 1m^3 / m^3 \quad (25)$$

Due to excess air, the volume of water

$$V_{c.} = V_{c.g.} + V_{H_2O} \quad (26)$$

$$V_{w.} = 5,39 + 1 = 6,39 m^3 / m^3 \quad (27)$$

The maximum normal flame propagation velocity (m / s) in a gas-air mixture can be determined by the formula:

$$U^{max}_н = \frac{\Sigma C_m H_n \cdot U_{C_m H_n}}{\Sigma C_m H_n} \quad (28)$$

$$U^{max}_н = \frac{45 \cdot 0,37}{45} = 0,37 m / s \quad (29)$$

With the known content $N_2$ and $CO_2$, % vol., normal flame propagation speed $U^b_n$, m/s for ballasted gases, correct by expression:

$$U^b_n = U^{max}_n (1 - 0,01N_2 - 0,012CO_2) \quad (30)$$

$$U^b_n = 0,37(1 - 0,01 \cdot 1 - 0,012 \cdot 52) = 0,14 \text{ m/s} \quad (31)$$

It is known that the average value of the theoretically necessary volume of air for complete combustion of natural gas is about 9.85 m3, which, in comparison with the volume defined above, necessary for complete combustion of bio gas equal to 4.43 m3, in 2.2 times less. Thus, in gas burners for the combustion of bio gas of the composition examined, this ratio must be observed.

The purpose of the gas burners is to organize the combustion of fuel in order to ensure a predetermined economically feasible operating mode of the boiler unit.

The gas burner achieves this goal by:

– Provision of supply and mixing among themselves the required amount of fuel and air;

– organization of complete combustion of fuel within the boiler furnace space;

– combustion of fuel in a mode that allows to ensure the level of heat transfer required in technological conditions in the working space of the boiler furnace.

Consequently, the entire cycle that makes up the process of burning the fuel (mixing-burning-heat transfer) must be performed with the highest efficiency of useful action [4].

The method of mixing gas with air is the main classification feature of burner devices and is divided into three groups:

According to the main classification feature - the method of mixing gas with air, burners are divided into groups:

– complete pre-mixing of gas with air;

partial premixing of gas with air;

– without premixing.

Given the low content of CH4 in biogas, it will be expedient to use a vortex-type burner with forced air supply and partial premixing of gas and air.

The merits of vortex burners include:

– Ability to work on low-calorie gas;

– capacity control can be carried out in a wide range;

– the possibility of working in heated air.

To study the formation of a gas-air mixture, a computer model of a low-pressure vortex gas burner GGV-100 was developed [5]. The model is built using the Solid Works software complex, the gas dynamics calculations were performed by the Solid Works Flow Simulation module.

For the calculation, the following initial values of the parameters were chosen: dimensions correspond to the dimensions of the GGV-100 burner; gas flow at the burner inlet Gr = 0,027778 $m^3$ / s (100 $m^3$ / h);

air flow at the burner inlet $G_B$ = 0,123056 $m^3$ / s (443 $m^3$ / h); the roughness of the walls is $\Delta$ = 100 μm; initial pressure in the combustion chamber $P_O$ = 101325 Pa.

Figure 1 shows a computer model of the GGV-100 vortex burner.

As a result of the calculation, the patterns of the change in the velocity of the gas-air mixture in the longitudinal-vertical plane of the section of the gas burner-furnace section shown in Fig. 2 and 3.

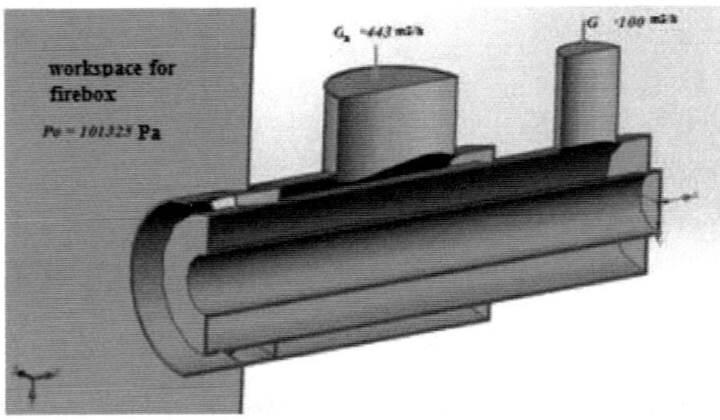


**Figure 1.** Vortex burner model

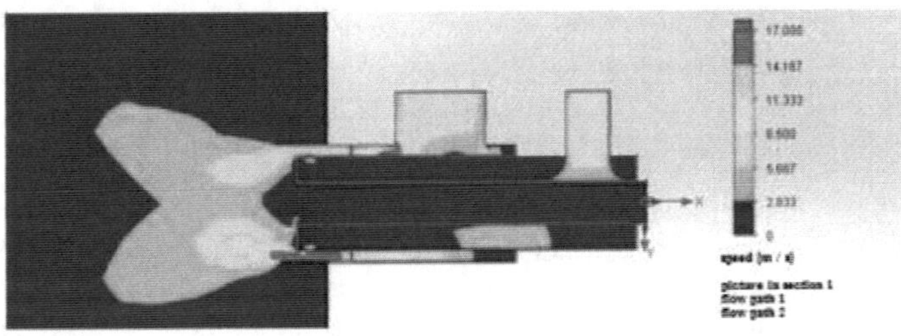

**Figure 2.** The is surface of the change in the speed of the gas-air mixture in the section of the section gas burner – furnace

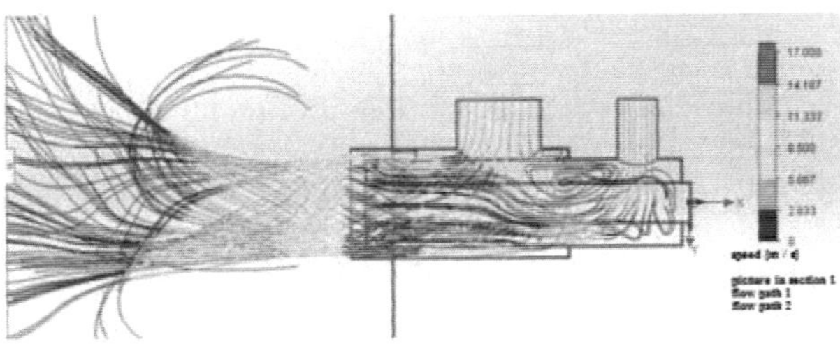

**Figure 3.** Isolines-trajectory changes in the speed of the gas-air mixture in the section of the section gas burner – furnace

Analyzing the data obtained on the graphs, one can draw the conclusion that the velocity of the gas-air flow at the exit from the burner to the to-mail space occurs at a speed ranging from 11 to 17 m / s. At the same time, for the kinetic principle of combustion used by this burner, the rate of the expiration of the gas-air mixture from it during the combustion of a low-calorie gas must be taken at least 10-15 m / s. This condition must be met in order to avoid flashing the flame into the burner [4]. However, due to the relatively short length of the site with the optimal speed, the use of bio gas in the gas burner under test requires a detailed maintenance of the gas-to-air balance regime.

Conclusions. As a result of the study, a model for the distribution of a gas-air mixture in a vortex gas burner was developed using the program module Solid works Flow Simulation, taking into account the release into the furnace space. The distribution of the velocity of the gas-air flow was determined and a comparative picture of the change in the value of the velocity of the gas-air flow at different sections of the burner-furnace assembly was constructed. On the basis of the computer-aided model of aerodynamics of the GVV-100 vortex burner, it is possible to further improve and modernize gas-burning devices of this type.

**Acknowledgments**
The article was prepared within development program of the Flagship Regional University on the basis of Belgorod State Technological University named after V.G. Shukhov.

**References**

[1] Suslov D 2014 Development of a gas supply system for an agricultural enterprise using biogas. *Vestnik BGTU named after V.G. Shukhov* vol 4 pp 183-186

[1] Komina G, Shkarovsky A, Marinenko E 2010 *Gas supply. Burning gases: Textbook* (VolgGASU) p124

[2] Timashev A, Berkunova T 2012 Determination of the moisture content of natural gases: Methodological instructions for conducting laboratory works (*Moscow: Russian State University of Oil and Gas named after I.M. Gubkina* ) p 23

[3] Sidelkovsky L, Yurenev V 1998 Boiler installations of industrial enterprises (*Moscow: Energoatomizdat* ) p 528

[4] Vintovkin A, Ladygichev M, Gusovsky V, Usachev A 2001 Modern burner devices (structures and technical characteristics) (*Moscow: Mashinostroenie- 1*) p 496